Environmental Engineering

Volume 7 Scientist and Science Series

Enders Anthony Robinson

Professor Emeritus of Applied Geophysics in the
Maurice Ewing and J. Lamar Worzel Chair
Columbia University in the City of New York

Goose Pond Press

Available from Amazon.com and other retail outlets

COVER. William Blake's picture of Sir Isaac Newton as a divine geometer (color print with pen & ink and watercolor, done in 1795)

Copyright © 2019

by

Enders A. Robinson

All Rights Reserved Worldwide

Goose Pond Press

Some scarce see nature at all.
But to the eyes of the man of imagination,
nature is imagination itself.
As a man is, so he sees.

—William Blake in his letter of August 16, 1799 to Rev. John Trusler

Dedicated to grandchildren

Chloe Alexandra Robinson

Bjorn Knowlton Robinson

Ashtyn Adelle Robinson

Contents

Chapter 1. Fukushima Nuclear Disaster   7

    Mirth and Melancholy   7
    Optimistic viewpoint   9
    Pessimistic viewpoint   10

Chapter 2. Environmental Engineering   13

    Environmental Earth Engineering   13
    Complexity and Simplicity   15
    Integrated Methodologies   19
    Raw Materials: Iron Ore, Aluminum, Copper, Nickel   23
    Depletion of Mineral and Petroleum Resources   33
    Research Framework   37

Chapter 3. Global Climate Change   42

    Global Change   42
    Carbon Dioxide and Global Warming   43
    Stratospheric Ozone Depletion and UV Radiation   50
    Volcanic Cooling   59
    Significant Variations of the Seasonal Climate   61
    Observing Global Change   65
    Reconciliation with Observations   67
    Predicting Global Change   68
    Seasonal to Inter-annual Forecasting   71
    Evaluating the consequences   72
    Mitigation Strategies   74
    Problems   75

Chapter 4. Air Pollution   79

    Pollutants in the Atmosphere   79
    Carbon monoxide (CO)   81
    Nitrogen oxides (NO_X)   85
    Hydrocarbons   85
    Atmospheric Pollution   86

Photochemical Reactions	87
Photochemical Oxidants	87
Indoor Air Pollution	92
Thermal Pollution and Noise	98

Chapter 5. Soils and solid waste — 100

Food production	100
Soils	103
Agriculture	105
Municipal Solid Waste	105
Disposal Options	107
Solid Waste Disposal	108
Resource Recovery	111
Hazardous Waste	115
Waste Processing and Handling	116
Resource Recovery Alternatives	118

Chapter 6. Water — 128

Hydrologic Cycle	128
Chemical composition	132
Drinking-Water Pollution and Human Health	135
Specific Hazards	138
Radioactive Contamination	146
Hazardous-Waste Sites and Groundwater Contamination	146
Pollution and Water Treatment	147

Chapter 7. Energy—Oil and Coal — 152

Petroleum	152
Non-renewable energy resources	157
Oil Exploration: Past and Future	161

Chapter 8. Nuclear Energy — 174

Units for Measuring Radiation	174
The Energy Perspective and Nuclear Power	176
The Chain Reaction	177

Breeder reactor	182
Sources of radiation	183
Radionuclides	188
Ionizing Radiation	189
Radiation doses by Canadian Nuclear Safety Commission	191
Effects of Radiation	193
Biological effects	195
Natural radiation	197
Chernobyl	198
Fukishima	203
Radioactive Waste	204
High-level waste (HLW)	207
Transuranic (TRU) waste	214
Low-level waste (LLW)	215
Environmental Impact of Nuclear Weapons	217
Other Sources of Radioactive Waste	221
Radioactive Waste Management	222
High-Level Radioactive Waste	224
Nuclear Testing	227
Plutonium	229
Fuel Reprocessing	233
Disposition of Excess Weapons Plutonium	240

Chapter 1. Fukushima Nuclear Disaster

"The eyes of the cheerful man and of the melancholy man are fixed upon the same creation; but very different are the aspects which it bears to them."—Albert Pike

Mirth and Melancholy

Following a major earthquake, a 15-metre tsunami disabled the power supply and cooling of three Fukushima Daiichi reactors, causing a nuclear accident on 11 March 2011. The Tokyo Electric Power Company (TEPCO) is currently responsible for the clean-up and decommissioning process at the Fukushima Daiichi site and in the surrounding exclusion zone. The plant had four reactor units. Units 1, 2-3 were in operation; unit 4 was shut down. The cores in units 1, 2, 3 largely melted in the first three days. The accident was rated 7 on the INES scale, due to high radioactive releases over days 4 to 6. All four reactors were written off due to damage in the accident. After two weeks, the three reactors (units 1, 2, 3) were stable with water addition and by July they were being cooled with recycled water from the new treatment plant. Official 'cold shutdown condition' was announced in mid-December. Apart from cooling, the basic ongoing task was to prevent release of radioactive materials, particularly in contaminated water leaked from the three units. This task became newsworthy in August 2013. There were no immediate deaths from radiation sickness. More than 100,000 people were evacuated from their homes. The government kept delaying the return of many of the evacuees. Official figures show that there have been well over 1000 deaths from maintaining the evacuation.

On 12 April 2011 nuclear regulators elevated the severity level of the nuclear emergency from 5 to 7—the highest level on the scale created by the International Atomic Energy Agency, placing it in the same category as the Chernobyl Nuclear Power Plant, near the city of Pripyat in the north of Ukraine in 1986. It was not until the middle of December 2011 that the facility was declared the facility stable, after the cold shutdown of the reactors was completed.

A second, but smaller, nuclear accident took place in August 2013 when approximately 300 tonnes (330 tons) of irradiated water used in ongoing cooling operations in reactors 1, 2, and 3 was discharged into the landscape surrounding the Fukushima Daiichi facility. The leak was the result of an open valve in the short barrier wall that surrounded several of the tanks used in radioactive water storage. The leak was severe enough to prompt Japan's Nuclear Regulation Authority to classify it as a level-3 nuclear incident.

Beginning in July 2013, evacuation orders were lifted in some areas characterized by lower levels of radiation both within and beyond the 20-km evacuation warning zone. By March 2017 all evacuation orders in the areas outside the so-called "difficult-to-return zone" had been lifted.

As of 2019 there are two divergent points of view on the current status of the Fukushima nuclear accident; namely, the **optimistic** and the **pessimistic**. The optimistic viewpoint is reminiscent of *L'Allegro*, a poem by John Milton published in his 1645. On the other hand the pessimistic viewpoint is reminiscent of *L'Allegro*, the companion poem of John Milton.

L'Allegro (which means "the happy man" in Italian) is invariably paired with the contrasting pastoral poem, *Il Penseroso* ("the melancholy man or serious man). *L'Allegro* invokes Mirth and other allegorical figures of joy and merriment, and extols the active and cheerful life. The poem ends with the two lines:

> These delights, if thou canst give,
> ***Mirth*** with thee, I mean to live.

Il Penseroso dwells on the pleasures of solitude and renunciation of the world. The poem ends with the two lines:

> These pleasures, ***Melancholy***, give,
> And I with thee will choose to live.

As to nuclear, should there be mirth or should there be melancholy?

Optimistic viewpoint

Most of Japan's power plants shut in the wake of the Fukushima nuclear disaster. But in 2015 the Prime Minister announced plans to restart reactors because the economy needed cheap energy and using fossil fuels risked huge carbon emission fines. It is aimed to have at least 12 in use by 2025.

The reactor of the Fukushima Daiichi nuclear power plant irrigation is continued to be poured water in order to maintain the current state of cold shutdown. The core fuels solidified after the melt down from Unit 1 to Unit 3, the spent fuels stored in the each storage pool and more than 1500 of the fuel rods which had been stored in the fuel storage pool of Unit 4 must be continued to cool down. No. 4 reactor did not have any fuels in the reactor due to the regular inspections at the time of the accident. However, even those spelt fuels generate heat even in the pool, so it become dangerous quickly if not continue cooling the spent fuel rods.

The work towards the decommissioning has been accelerated by conducting rubble removal at the site concurrently with cooling down. The plan of the dismantling of the facility will be terminated over 30-40 years, but the decommissioning work has been hampered by difficulties in the very high radioactivity. In June 2012, the dosimeter was inserted by using the endoscope from the reactor building of Unit 1 to the first floor of the basement. In the basement, the level of the radiation recorded the maximum hourly 10.3 Sv in 5 meter water depth. This was the figure that means the high probability of the death if a person exposes for one hour. The only work which can be started in the beginning is the removal of fuel rods from the fuel pool in each unit. There is no prospect about how to eject the melted down fuels in 1, 2,-3 units.

The radioactive materials from the Fukushima Daiichi nuclear power plant are still leaking. Of course, the amount is small compared with at the time of the accident. However, it should be recognized as a "radiation leak accident" if it is in the normal case. Most of radioactive materials are leaking from units 1, 2,-3 because of lack of the confinement function at the collapsed reactor building from units

1, 2, 3. Still Release of maximum 10 million Becquerel's per hour is continuing. 400 tons of groundwater per day was flowed into the reactor building and acuminated as contaminated water. TEPCO has created a large amount of on-site storage tank to store the contaminated water. By November 2012, the well excavation begins in order to pump up groundwater before it is contaminated and flow into the sea. In the future, the contained water will be purified up to the level of no detection of radiation by the special equipment and release to the sea, according to TEPCO.

Japan seems to have been very successful in controlling radioactivity in foodstuffs to date, but still there is much radioactivity in the environment around the reactor sites and in the regions that were evacuated. But while there are safety concerns, the local government are at pains to stress they are making progress in recovering and revitalising the Fukushima Prefecture's affected areas. Officials also point out that only about 3 percent of what is Japan's third largest area was in the danger zone which was near the coast and that the rest of the region has been unfairly tainted.

Pessimistic viewpoint

Uranium and other radioactive materials, such as cesium and technetium, have been found in tiny particles released from the damaged Fukushima Daiichi nuclear reactors. This could mean the environmental impact from the fallout may last much longer than previously expected. For the first time, the fallout of Fukushima Daiichi nuclear reactor fuel debris into the surrounding environment has been "explicitly revealed" by the study.

Scientists have been looking at extremely small pieces of debris, known as micro-particles, which were released into the environment during the initial disaster in 2011. The researchers discovered uranium from nuclear fuel embedded in or associated with cesium-rich micro particles that were emitted from the plant's reactors during the meltdowns. The particles found measure just five micrometres or less; approximately 20 times smaller than the width of a human hair. The size of the particles means humans could inhale them. The reactor debris fragments were found inside the nuclear exclusion zone, in paddy soils and at an abandoned aquaculture center located several kilometres from the nuclear plant.

It was previously thought that only volatile, gaseous radionuclides such as cesium and iodine were released from the damaged reactors. Now it is becoming clear that small, solid particles were also emitted, and that some of these particles contain very long-lived radionuclides; for example, uranium has a half-life of billions of years. Research strongly suggests there is a need for further detailed investigation on Fukushima fuel debris, inside, and potentially outside the nuclear exclusion zone. Although it is extremely difficult to get samples from such an inhospitable environment, further work will enhance our understanding of the long-term behavior of the fuel debris nano-particles and their impact. Better knowledge of the released microparticles is important as it provides much needed data on the status of the melted nuclear fuels in the damaged reactors. This would provide extremely useful information for TEPCO's decommissioning strategy. At present, chemical data on the fuel debris located within the damaged nuclear reactors is impossible to get due to the high levels of radiation. The microparticles found by the international team of researchers will provide vital clues on the decommissioning challenges that lie ahead.

It is not over; it is just getting started as the radiation flowing from the plant is increasing. The half-life of the various radioactive elements ensures that the radiation levels will just keep rising. This appears to be an "extinction level event" of the Pacific, perhaps much more. When the floor in which the rods are stored finally collapses and the resulting accelerated decay reaches critical mass, the nuclear explosion will spread radioactivity across the northern hemisphere, The radiation will take a time to take its toll. It is a silent, invisible killer that has time on its side. One way or the other, the damage has been done. It is known that nuclear radiation can take years to produce cancer killing effects.

Japan is only in year eight of an accident that will continue to threaten public health, and the environment well into the twenty-first century. If the government can create the illusion of the region that that has recovered from the nuclear accident they think it will reduce public opposition. But meanwhile the crisis continues at the Fukushima plant. It is said that the massive ice wall built at the nuclear plant to stop contamination of groundwater is a symbol of this failure and deception. The reality of a nuclear disaster knows no end. Today there are areas of Fukushima where radiation levels could give a person's maximum annual recommended dose within a week. Of particular concern with regards to poorly paid decontamination workers, thousands of whom have been involved in attempts to decontaminate radiation around people's homes, along roads and in narrow strips of forest. The government claims decontamination has been completed in 100 percent of affected areas after a ten billion dollar clean-up operation. They do not take into account that 70-80 percent of two of the most contaminated districts are forested mountains, which are impossible to decontaminate. In areas opened in March 2017 for people to return have radiation levels that pose a risk until 2050"These areas are still too high in radiation for people to return safely and is one reason so few people are returning. Meanwhile heavy-handed tactics are being used with some fearful residents reporting that they have been warned they will not receive lifeline compensation cash if they don't comply. The plant and the areas remain a danger zone for humans. It seems there is no end in sight for the release of radioactive water from the site and these releases will inevitably put more radioactivity into the food chain if the local waters are fished.

Chapter 2. Environmental Earth Engineering

"Man is not born to solve the problem of the universe, but to find out what he has to do; and to restrain himself within the limits of his comprehension."— Johann Wolfgang von Goethe

Engineering Methods for Environmental Protection

Environmental issues must be addressed. Some of the most important are environmental degradation, atmospheric changes, depletion of resources, contamination of water supplies, soil erosion, and destruction of species. From one point of view, to solve or mitigate these problems, the efforts must be compartmentalized and narrowly focused. They must be attacked from the perspective of the pertinent discipline. A problem such as the emission of carbon dioxide into the atmosphere has to be treated in the realm of atmospheric science. From another point of view the Earth must be studied as a system of linked components and processes. Only in this way can we find effective solutions to the environmental and resource problems. We must access the scope, goals, and emerging research issues. The process of understanding the Earth is in its initial state. For humans to survive and prosper over the long run, it is necessary to understand the intricacies of interacting earth systems. In extracting resources from the earth and discarding those resources back into the earth, humans exert a major influence on the geological processes and cycles. To understand the rapid environmental changes resulting from human activities, it is necessary form a picture of the environment from the recent past all the way back through geological time. A comprehensive environmental picture is evolving and expanding the frontiers of knowledge. Interrelations of the physical, chemical, geological, and biological processes that characterize the Earth and its history are being discovered.

ENVIRONMENTAL EARTH ENGINEERING is the study of engineering methods to correct environmental problems affecting the earth system. It is revolutionizing the understanding and correction of many Earth problems. Environmental Earth engineering includes questions about what makes the Earth work. It can help us solve problems such as how natural systems operate, but it is more. It is a crucial component in efforts to solve mankind's current environmental problems A visionary and broad-ranging study of our planet would have four crucial aspects: (1) environmental observations, including those based on space technologies and international collaborations; (2) development and application of new instruments; (3) utilization of new computer technologies; and (4) support of advanced training in science and technology. In addition to conventional courses, the study of environmental Earth engineering would include:

1. Courses in earth system science

2. Courses in such areas as hydrology, land use, engineering, geology, environmental and urban geology, and waste disposal

3. Courses that explore educational opportunities that bridge the needs of earth science and engineering departments.

Scientists reconstruct worldwide snapshots of environmental systems in order to test the effectiveness of engineering models. These models simulate actual conditions and thereby improve the ability to predict the impacts of future changes in the environmental system. Evidence from subsurface geology, rivers, lake and ocean sediments, and land use document that the environment may have changed frequently and abruptly over the last few years. Additionally, data from time-series reveal previously unknown process relationships that help establish the rate of change within the system.

Environmental research is directed toward understanding shorter time scale geological processes, such as hydrological circulation. This research reveals a wide variety of important effects in overlying water columns, including formation of turbulent, buoyant plumes, the introduction of chemicals and biota, and the modification of circulation and mixing patterns.

Chapter 2. Environmental Earth Engineering

Complexity and Simplicity

The different attitudes of environmental scientists and the traditional scientists as to the nature of the world illustrates their different perspective. This different perspective has a strong influence upon their views of and the importance they attach to their approaches. A traditional scientist looks for simple, symmetrical, and elegant ways to describe the world. He seeks to reduce the diverse materials of nature to a handful of building-blocks obeying elementary laws governing the operation of nature. Yet when we examine the workaday world it is nothing of the sort. Our daily lives, the workings of our businesses, national economies, local ecologies, or weather systems are anything but simple. Rather, they are a mixture of complexity. They are governed by a assembly of interlinked processes that possess neither symmetry nor elegance. As complexity becomes more organized so the range of phenomena that issue from it grow with unpredictable subtlety. Indeed, an environmentalist would not talk about simplicity and symmetry, but would describe the interlinked complexity of the outcomes of the natural selection under assault. Neither planned nor guided, there is no reason for the changes on the environment to be simple; there are many more ways for them to be complicated. Their primary characteristic is persistence, or stability, rather than simplicity.

So how are we to address this apparent dichotomy: Is the world simple or is it complicated? The answer is important because the traditional scientist, impressed by the simplicity of the world, will find it neatly encapsulated by descriptions of pattern and structure. The environmental scientist will find all his problems too hard for traditional methods to work. He will need to model, to approximate, and to guess in order to quantify what he sees. He will ascribe no elevated status to mathematics alone; instead he will be persuaded that the world cannot be captured by its simple structures.

There seems to be two different ways of thinking about what is important for our understanding of the world. One approach is to emphasize the timeless and unchanging aspects of the world as being the most fundamental. These are the invariant blueprints of which all observed things are the examples. The observed happenings are therefore less fundamental than the unchanging blueprints that govern them. This emphasis tends to de-emphasize the process of change and the notion of time from the description of things. This desire can be realized to a surprising extent. The traditional laws of Nature dictating change can always be recast as equivalent statements that certain quantities remain unchanged. The replacement of laws of Nature governing changes in space and time by statements that certain quantities remain unchanging, these statements of solidarity can in turn be replaced by the dicta that certain patterns or 'symmetries' be preserved in Nature. The requirement that such powerful invariances be preserved turns out to demand the existence of the forces of Nature and to dictate the way in which particles interact with each other.

The known forces of Nature are founded upon the immutability of some pattern when changes occur. But there is a second tradition in the study of Nature that, until recently, has been less popular than the search for the invariants of Nature. This perspective lays emphasis upon the observable happenings in the world rather than the unobservable invariants behind it. As a result the process of temporal change is regarded as fundamental. The advocates of such an emphasis draw their intuition from the study of living things. This approach the world is complicated and messy and its aspects cannot be entirely explained by simple laws acting behind the scenes.

To understand the real difference between the simple view of the world and the complicated perspective we need to appreciate one important fact about the world: symmetrical laws of Nature need not have outcomes which possess the same symmetries as those laws. Outcomes are much more complicated things than laws of Nature themselves. Moreover, we do not observe the laws of Nature: we observe only the outcomes of those laws and from the heap of broken symmetries before us we must work backwards to reconstruct the pristine laws behind the appearances. Sometimes this is very easy to do, but often it is impracticable because of the sensitivity of the direction of the symmetry-breaking to the whims of

Chapter 2. Environmental Earth Engineering

the environment. But we have learnt one important lesson. This process of symmetry-breaking explains how we can reconcile the existence of observed complexity with underlying laws of Nature that are simple.

The life scientist, or the economist, troubles himself not with 'laws' of Nature. The focus there is entirely upon the complicated outcomes of the underlying laws. Contrary to pristine symmetry and mathematical beauty, the complicated results of the real world possess neither of those desirable features. Instead, we are faced with understanding outcomes that are separated from the underlying 'simple' laws of physics by a long sequence of hidden symmetry-breakings. The world can be both simple and complicated in important ways and the aspect that impresses you most will depend upon whether you are more concerned with the laws of Nature or their outcomes. In environmental engineering, we must realign our focus of attention upon the outcomes, having come to appreciate that there exist sequences of events which cannot be replaced by timeless invariants in the traditional scientific manner.

The study of complexity in the abstract deals with nature as a general phenomenon not necessarily tied to a particular complicated physical situation. These are ideas that belong in a discussion of randomness. The lack of some abbreviated representation means that there exists no symmetry or invariance whose simple preservation is equivalent to the data content of the sequence. For that would be a compression of the sequence. The outcomes contain a level of complexity that requires nothing less than their explicit listing to capture their full information content.

Besides elevating the study of outcomes to something that is not necessarily included within the study of natural laws this notion of compressibility gives simple ways of characterizing many of our intellectual activities. Environmental science must be involved with the search for compressions: the laws of Nature are the compressions of our sense data. Moreover, the apparent success of this process hinges upon two superficial features of things: the physical world that we observe seems to be surprisingly amenable to compression, and the brain is remarkably good at effecting compressions when presented with events. In a predominantly incompressible world we would not have scientists but archivists who simply recorded every observed event. The compressibility of many aspects of the world saves us. We can use a simple law of motion to describe the motion of heavenly bodies instead of having to keep a record of their positions and velocities at all times. Yet, clearly, this compressibility and the brain's remarkable ability to make sense of complicated things is an important necessary condition for our own existence. We could not survive as 'intelligent' observers and readers of a book like this in a world where no compressions were possible or with brains that produced imaginary or erroneous compressions. A certain level of predictability and innate predictive power is required for the successful evolution and survival of living things. It has clearly proved efficacious to overdevelop our pattern-recognition capability (presumably because if you see tigers in the bushes when there are none your friends will merely call you paranoiac, whereas if you fail to see tigers in the bushes when there are, then your continued survival must be rather doubtful). As a result we see canals on Mars and all manner of exotic things lurking in inkblots. Yet, the brain cannot gather all the information potentially on offer to it; that would be as impractical as gathering none—would we really want to receive information about every last electron orbital when we looked at a painting? It overcomes this problem by storing only a part of all the information available to the senses. Our physiological make-up helps to effect this truncation by placing limits on the intensities of light and sound that we can respond to. However, this serves to warn us that the brain would effect a compression of the observed information even if one did not truly exist. Furthermore, we know that many aspects of the scientific enterprise set out to truncate the information available to us in order to effect a compression, for example by random sampling to obtain a representative opinion poll.

In the so-called hard sciences the most important characteristic of their subject matter, that encourages compression, is the existence of simple idealizations of complicated situations which can underpin very accurate approximations to the true state of affairs. Subsequently, the idealizations can be relaxed slightly and one can proceed step by step towards a more realistic description that allows for the presence of small differences, then to further realism, and so forth. By contrast, many of the 'soft' sciences which seek to apply mathematics to such things as social behavior, prison riots, or psychological responses fail to produce a significant body of sure knowledge because their subject matter is far less compressible and does not readily provide obvious and useful idealizations from which one can proceed towards better and better approximations to reality. Different types of scientist focus upon different aspects of the physical world and as a result will not be equally impressed by the role played by the other.

Integrated Methodologies

Integrated assessments bring research results from natural, social, and policy sciences into a framework that helps decision makers identify and evaluate actions to respond to environmental changes. Integrated assessments can help set priorities for the natural as well as the social and policy science research areas by providing information about the relative value to decision makers of information that is likely to result from research investments. Another major objective of integrated assessment is to develop information about interactions of complex, linked systems and problems and to make this information available in the decision-making process. Integrated assessments can advance the state of research by providing a framework for synthesizing findings and data from the physical, chemical, geological, biological, economic, social, and health sciences.

It is proposed to markedly increase its support of integrated assessment activities. Some of the increase will support the development of methods and tools. One set of tools, integrated assessment models (IAMs), are an emerging type of model that link **human forcing functions** (e.g., greenhouse gas emissions), the effects of those forcing on the Earth-system, the resulting impacts on humans and ecosystems, and the economic and environmental consequences of potential responses. The models allow different environmental climate change policy options to be evaluated in terms of their influence on various parameters, such as their effect on Gross Domestic Product or employment. A number of the climate change IAMs have evolved from cost models developed primarily in the 1970s and 1980s that predicted emissions and calculated the cost of meeting various emission or concentration goals.

IAMs can be of various types, depending on the issue being addressed. Some IAMs are environmentally aggregated and do not, for example, distinguish the United States from the rest of the world. Such models can be used to study sequential decisions and the value of additional information. These models cannot, however, facilitate evaluation of policy instruments such as appliance efficiency standards, automobile efficiency standards, or having separate taxes on different fuels. IAMs can also be spatially disaggregated, dividing the world economy into up to a few dozen regions. These larger models include more realism in their treatment of specific aspects, such as regional emission patterns and economic activity by sector. This type of IAM will be better suited to answer policy questions such as how to trade off the different greenhouse gases using economic as well as natural science criteria. For example, a biomass energy option or carbon sequestration program may become less effective if climate change is harmful to forest systems. Similarly, the model may indicate that international application of policies such as are required in the United States by the Clean Air Act may induce changes in future climate (e.g. through requirements for reduced emissions of particulate and sulfur) that need to be considered in evaluations of climate changes due to greenhouse gas emissions policies.

Phenomena associated with environmental change can affect currently stable regional environments and indigenous natural resources, and could have profound implications for land-use management. Regional-scale integrated assessment models are being developed based on detailed modeling of physical, chemical, geological, biological, and human systems. In many cases these

Chapter 2. Environmental Earth Engineering

models can define critical thresholds of environmental change. Early work is proceeding on such models for regions including the Great Lakes, the Great Plains, and the Southwest.

Support for integrated assessment methods will enhance efforts to develop reduced-form models that mimic the more complete natural science process models. The program will also encourage examination of a variety of methodological issues related to integrated modeling, including development of additional quantitative and qualitative approaches for integration. Related research will include: (a) modeling and evaluating the implications of rates of environmental change in addition to the extent of absolute change; (b) evaluating the impact of an increased frequency of extreme events; (c) uncertainty analysis of physical, economic, and biological parameters; (d) review of the successes and shortcomings of past integrated assessment studies (e.g. acid precipitation, stratospheric ozone, supersonic transport); and (e) comparison and evaluation of integrated assessment methodologies.

Improvements in integrated assessment capabilities require investments in both integrated models and in the research that provides the foundation on which models are based. Integrated models that can be used to conduct comprehensive assessments and evaluate policies and other options require adaptation of reduced-form models, which are already under development for many natural systems. They also will require development of accounting mechanisms that will enable non-monetary factors to be included. New insights into integrated assessment modeling are resulting from evaluating modeling approaches that have already been developed and tested.

Various sectoral analyses are planned to evaluate the consequences of environmental engineering. There is a need for increased understanding of the ability for communities to adapt to new policies that may be imposed to address the environmental issues. Research on how consumption patterns may change with a given method is necessary in developing response options. For example, options for responding to possible energy savings in buildings must take into account the reduced demand for space-heating with the increased demand for space-cooling.

Uncertainty about the adaptability of communities to environmental actions is due in part to uncertainty about future technological change. How will industrial development, future industrial and consumer products, and technological improvements change polluting emissions, help abatement of, and lead to adaptation to environmental changes? How will the supply and demand for goods and services change? Answers to these questions have implications both for predicting future changes and for evaluating the effectiveness of national and international government policies to reduce the impacts of change. Technological innovations and changing production patterns could greatly ameliorate or exacerbate both abatement and adaptation. Research on the innovation process and the subsequent diffusion of new technologies and products will be expanded to address these issues.

An important area of research concerns resource use and management related to environmental earth engineering. This line of research directly supports the pursuit of sustainability. Resource issues are location and scale dependent, and numerous advances have been made in knowledge of the ways that resources are used and managed in local and regional settings. Many advances have resulted from more sophisticated use of remotely sensed data. Important improvements have been made in the interpretation of satellite imagery for different areas where tropical rainforests abound, for example. When coupled with field-based studies of land-use and forest-regeneration processes, more complete understanding has been achieved of the successive stages of vegetation on sites where trees have been cleared or on which biomass burning has occurred. This research has shed light on natural, demographic, economic, and social factors that result in much more rapid regeneration in some locales than in others. This kind of knowledge is useful for designing more effective land use policies.

Chapter 2. Environmental Earth Engineering

There is a feedback effect between human activities and environmental change. In other words, human activities can greatly influence environmental change and also human activities can be significantly influenced by environmental change. Changes in land use can be a major driving force of environmental change. Conversion of natural forests into croplands and human settlements, overgrazing of grasslands, intensive planting and fertilization of agricultural lands can have significant impacts on the environmental carbon budget and fluxes of greenhouse gases. For example we can address questions such as:

(1) How has land cover been changed by human use over the last 300 years?

(2) What are the major human causes of land-use change in different geographical and historical contexts?

(3) How might these land-use changes affect land cover in the next 50 years?

(4) How might changes in climate and environmental biogeochemistry affect both land use and land cover?

Raw Materials: Iron Ore, Aluminum, Copper, Nickel

Oil is a form of energy and its availability is indispensable to running modern industries. Other minerals are also of immense importance: iron, copper, aluminum, chromium, cobalt, gold, titanium, tungsten, platinum, diamonds, uranium. These materials come from naturally occurring minerals. The Earth contains enormous reserves but often the mineral deposits that are economically exploitable are concentrated in specific regions of the world. The industrialized countries require large quantities of those raw materials and must import much of their supply from other places. The concentration of global oil resources in the Middle East illustrate that oil is the most critical material in the world. In recent years most of those minerals have been the subject of violent price and supply fluctuations often more unstable than the oil markets.

On a national level each country has a different raw materials demand and supply framework, although fuels and the so-called basic materials such as steel, copper, aluminum, nickel, lead, zinc, and tin are universally consumed in varying quantities by all countries. Variou factors determine which materials are considered strategic to the nation as a whole and which are only critical to the existence of specific industries. Several factors influence the supply and consumption of various raw materials in any particular country. These include the existence of mineral deposits, availability of capital and technology, sufficient energy supplies, transportation, infrastructure, and industrial demand. In some countries, the mining and production of raw materials is the major industry. Exports of raw materials provide foreign exchange that is required to pay for imports of equipment and technology to keep the economy in operation.

If there were the possibility of a severe shortage or outright supply disruption of a material that cannot be readily substituted in an industry, then it must be considered as being strategic or critical to the end-users. This is particularly true if such a material originates from foreign sources outside the political control of the country in which end-users and their industries are located. Oil is an obvious example, particularly in the case of countries that do not possess any domestic resources of this mineral. Unexpected price escalations may also play havoc with production schedules in cases when the end product contains a large proportion of materials whose prices increase significantly. On the other hand even drastic price increases of materials that are used in very small quantities in some products may not be critical to end-users. In such cases availability of such indispensable materials and unexpected shortages regardless of price may nevertheless create severe problems to the end-user industries.

Numerous danger points signal the criticality of each particular material to end-users that can be determined in advance. These range from the number and location of supply sources to various characteristics of the material that may threaten excessive

Chapter 2. Environmental Earth Engineering 25

regulation of its production and its eventual disappearance from the market. The existence of several danger characteristics with regard to a particular material does not in itself imply that the material's supplies will be inadequate in the future. Whether these danger points will have a disruptive effect on the supplies depends on other events and conditions of a political, economic, and social nature. Most of these are completely beyond the control of the end-user regardless of whether the materials originate from domestic or foreign sources. Many such events are also beyond the control of materials traders and distributors who are the intermediaries between the actual producer and the end-users and whose business objectives are often at variance with those of the materials-consuming industries.

Those trigger events and conditions are often political in nature but are not necessarily unpredictable. Materials managers whose enterprises depend on strategic and critical materials that exhibit many danger characteristics would do well to become familiar with the sources of their materials and the political climates of those environments. They can protect their organizations from serious loss of revenues or work disruptions by assessing political risks involved well in advance and prepare alternative action plans. In a world of materials shortages this course of action is the *sine qua non* of survival. The events and conditions that must be monitored for this purpose both on the domestic scene and in foreign countries include labor costs, strikes, civil unrest, wars, cartels, boycotts, embargoes, terrorist activities, revolutions, new legislation, taxation, as well as acts of God in distant places such as droughts, floods, fires, and earthquakes.

Oil provides the most convenient source of energy. However these are many other sources of energy. They include natural gas, coal, lignite, and biogas, as well as hydroelectric, nuclear, geothermal, solar, tidal, and wind power. Nuclear power is well advanced in some countries. That, however, creates another dependence on sources of uranium for nuclear fuels, making that metal a strategic energy resource. In the process of developing nuclear power, there is the added factor of nuclear weapons development. This makes uranium doubly strategic as a raw material to most countries in the world. It is also necessary to explore the problems associated with possible shortages or supply disruptions of critical non-energy resources such as metals and minerals that are also vital to the operation of modern economies.

Availability of adequate energy supplies is also of great significance to the minerals and metals production cycle and must be taken into account. Energy in one form or another is indispensable in exploration, mining, processing, refining, and transportation of raw materials. The different metals in their finished form can indeed be rated according to the amount of energy that is required to produce a unit amount of each. High cost or shortages of energy in a particular country or region will increasingly play a role in determining which metals will substitute for others and where production may be discontinued. In the United States, for example, 4 percent of all the energy consumed is used to produce aluminum alone. But recovery of aluminum from scrap can be accomplished with only a fraction of that energy; therefore as prices of energy continue to escalate organizations that are heavy end-users of aluminum may find it profitable to engage in large scrap recovery operations.

The Persian Gulf and its coastal areas are the world's largest single source of petroleum. Energy-rich countries such as Bahrain in the Persian Gulf have developed aluminum refining capacity in their own territory based on cheap and abundant local energy supplies. Such developments may further reduce domestic aluminum production capacity in other countries where energy is much more expensive and may lead to strange new arrangements between countries that have the energy and those with deposits of strategic minerals. There is a discernible trend among energy-rich states toward the development of energy intensive industries in their own territories. But if these countries do not possess sufficiently large markets to absorb the output of such industries there is the danger of disruption of existing markets in other countries through exports of products at significantly lower prices to capture market shares. Interestingly the centrally planned economies can protect themselves by strict foreign trade monopolies under central political control, but the free market economies are vulnerable to domestic output declines and unemployment in several threatened industries. Steel and textile industries are good examples of that type of industrial setback.

The cost of energy represents about 70 percent of the total value of the most valuable minerals. About 25 percent constitutes metallic minerals. The remaining 75 percent are minerals such as diamonds, salt, phosphates, asbestos, sulfur, mica, fluorspar, graphite, and asphalt. Several of the metallic minerals are produced in great quantities in the form of ores from which the basic metals in most common usage are extracted. These include iron ore, copper, aluminum, nickel, tin, zinc, and lead. Production of some of the metals, including iron, copper, aluminum, zinc, and lead, reaches millions of tons per year on a global basis. These basic metals are critical to many industries because they are used in such large quantities.

Aluminum, in the form of bauxite, and iron ore are respectively the third and the fourth most abundant elements in the earth's crust. The other basic metals are considerably less abundant, and together with all other elements constitute only a few percent of the earth's crust. Nevertheless the volume of the earth is so huge that millions of tons of all those metals can be produced every year. Despite the fact that aluminum and iron ore are so abundant there still exists the possibility of a producer's cartel because these metals are very widely used and because their ore deposits are very uneven and the highest grade areas are concentrated in only a few countries.

Iron Ore
This mineral is the primary source of iron and steel, basic to any industrial system. Major use in construction, shipbuilding, railroads, machine tools, automobiles, heavy equipment, agricultural machinery, and numerous consumer appliances makes this metal the most common in the world. However, over 31 percent of high-grade iron ore reserves is believed to be located in the Russia, which is by

far also the largest producer of iron ore in the world, accounting for almost a third of the total. Another 45 percent of iron ore reserves is located in Brazil, Canada, Australia, and India. The United States is believed to contain only about 6 percent of global reserves and is a net importer of iron ore from abroad. China is actually the second largest producer of iron ore in the world after the Russia, but the quality of Chinese ores is low. China is also a major importer of some ores from Australia and steels from Japan, West Germany, and other countries. The resulting steels that are produced from iron ores in the steel making process are widely used for industrial and military applications. High quality steels are alloys that also require large quantities of manganese, chromium, nickel, molybdenum, vanadium, and smaller amounts of other metals.

Aluminum
The use of aluminum in the world now exceeds the use of any other metal except iron. Construction industries are the largest consumers of aluminum, and transportation, packaging, electrical, and telecommunications industries are also major users of the metal. Transportation use includes aircraft manufacturing, for which aluminum is critical. Civilian jet transports and helicopters are also important in the export trade of major aircraft manufacturing countries, including the United States, France, the United Kingdom, Germany, and Russia. Military aircraft are major weapons systems and also contribute to the export trade of all major aircraft manufacturing countries. These applications make aluminum a very strategic material, particularly to those countries that do not possess bauxite deposits within their territories.

Bauxite is a sedimentary rock with relatively high aluminum content. It is the world's main source of aluminum. Bauxite consists mostly of the aluminum minerals gibbsite, boehmite, and diaspore, mixed with the two iron oxides goethite and hematite, the aluminum clay mineral kaolinite and small amounts of anatase and ilmenite.

Geographically major high-grade bauxite reserves are located in Australia, Guinea, Brazil, and Jamaica that together account for 75 percent of all known reserves. Alumina is a white solid, aluminum

Chapter 2. Environmental Earth Engineering

oxide that occurs in bauxite and is found in crystalline form as the main constituent of corundum, sapphire, and other minerals. The bauxites are subjected to hydrometallurgical processing to produce alumina that in turn is reduced to aluminum by an electrolytic process which consumes very large quantities of energy in the form of electric power. Although aluminum may be substituted by magnesium, production of that metal requires even larger quantities of energy. The United States is the largest producer of primary aluminum metal from imported bauxite and alumina. Russia is the second largest aluminum producer from its own domestic bauxites, and it also imports some aluminum from other producers. Japan, Canada, West Germany, and Norway are the next largest aluminum-producing countries.

Copper
After oil, coal, and natural gas, copper production accounts for over 6 percent of the value of all the minerals produced in the world. Copper is one of the first metals known and used by mankind, and the importance of this metal lies in its excellent characteristics as a conductor of heat and electricity. Copper is indispensable for use in electric power transmission, communications, generators, motors, transformers, switch gear, heat exchangers, condensers, air conditioning, refrigeration, tubing and piping, bearings, and military cartridge and shell casings. It is one of the most strategic materials and can substitute for many other metals, although it is itself hard to replace.

The United States and Chile are estimated to contain 40 percent of the world's copper reserves. Canada, Russia, Peru, and Zambia are believed to account for another 32 percent of the total. However four developing countries—namely, Chile, Peru, Zambia, and Zaire—account for over 80 percent of copper exports in the world. Poland is another country that recently developed its significant copper deposits. Copper is the third most important raw material moving in foreign trade after oil and wheat and accounts for about 3 percent of global exports of all primary products.

Nickel

This metal is one of the most versatile alloying materials used in the production of stainless steels and corrosion resistant super alloys capable of withstanding very high temperatures. Nickel can be substituted in almost every use but only at increased cost or loss of product performance. The metal has numerous strategic applications and is essential in military uses, including nuclear applications, jet engines, aircraft frames, submarines, armor plating, gun barrels, and rocket motor casings. About 90 percent of nickel is consumed in the form of alloys; the remaining consumption consists of use in electroplating, electrical equipment, machinery, vehicles, catalysts, batteries, fuel cells, ceramics, and household appliances.

As much as 25 percent of known nickel reserves of the world are in New Caledonia, a French Overseas Territory located in the South Pacific northeast of Australia. Canada is believed to account for 15 percent and Russia for 14 percent of world reserves. Indonesia has 13 percent; the Philippines and Australia, about 9 percent each; and Cuba, 6 percent of the total. Russia is now believed to be the largest producer of refined nickel.   Canada, Japan, Australia, New Caledonia (with France), and Cuba are the next largest refined nickel producers.

Tin
Tin is considered essential to an industrial society, and for many of this metal's applications there are no completely satisfactory substitutes. At least one-third of all the tin is used in the manufacture of cans and containers. Other important uses are in solders, bearing metals, brass and bronze, tinning, foils, ceramics, pigments, and miscellaneous chemical uses. About 20 percent of the tin used in the United States is reclaimed from scrap. Major tin reserves exist in South East Asian countries. Indonesia is estimated to contain 24 percent of world reserves; the People's Republic of China has 15 percent; Thailand, 12 percent; Bolivia, about 10 percent; Malaysia, over 8 percent; and Russia and Brazil, about 6 percent each. Nigeria, Australia, Zaire, and Burma also possess some tin. Malaysia is the leading primary tin producer in the world, accounting for about 40 percent of the total, followed by Russia, which produces about 17

percent. Thailand, Indonesia, Bolivia, and China are the next largest tin producers, ranging from 16 to 8 percent of the world's tin output.

Zinc
Zinc is the third most commonly used nonferrous metal in the world after copper and aluminum. It is versatile and essential in modern living. Principal use is in automobile die castings, typewriter chassis, housings, galvanizing iron and steel products, as alloying element in brass, in rubber, and as paint pigment. No adequate substitute is known for zinc in its use in galvanizing steel. Brass sheet as cartridge brass is used in large quantities to manufacture small arms ammunition shell casings for sporting and military use. Highest zinc reserves are in the United States and Canada, which are estimated to contain about 27 and 20 percent of global reserves, respectively. Other major zinc reserves are in Australia, Peru, Japan, and Russia, but zinc mining also is significant in Mexico, Zaire, Italy, Germany and Poland.

Russia is the largest producer of refined zinc, followed by Japan, Canada, and the United States. Poland, West Germany, Australia, Belgium, France, Italy, Spain, Finland, and Mexico are among the larger producers. In total at least 30 countries produce zinc, which means that many alternative sources of the metal exist. Despite this wide occurrence and production zinc is only the twenty-fourth most abundant metal in the earth's crust, but it is being used in relatively large quantities.

Lead
The principal use of lead is in storage batteries, bearings, gasoline antiknock additives, and in the electrical industries. Lead is also used in paints, solder, type metal, and some brasses and bronzes. More strategic uses include ammunition and as shielding against nuclear radioactivity. In terms of the amount of nonferrous metals consumed, lead ranks fourth after aluminum, copper, and zinc (it is often coproduced with zinc). The United States is estimated to contain 37 percent of global lead reserves, followed by Canada with 13 percent and Australia with 12 percent. Nevertheless in 1978 the Russia was the largest smelter lead producer, having just surpassed

the United States in 1977. Australia, Japan, Canada, Mexico, and Bulgaria are also major lead-producing countries. At least 22 countries are significant lead producers. Lead is one of the best nuclear radioactivity shielding materials.

Uranium
One of the most strategic and controversial raw materials in the world today is uranium, the nuclear fuels derived from uranium, and plutonium, a prime nuclear weapons material. A large number of countries have nuclear power plants to produce part of their energy requirements. At least 55 countries of the world have a total of about 800 nuclear reactors in operation, under construction, on order, or in planning stages to go into operation. In addition at least 18 countries are planning to operate their own uranium enrichment or plutonium separation plants. Governments and nuclear industries of many countries are committed to develop nuclear power. One of the attractions of nuclear power plants and nuclear fuel manufacturing capability is the potential for developing nuclear weapons. The very existence of a nonproliferation treaty only confirms this trend among governments big and small, and many political analysts are convinced that whereas nuclear weapons proliferation may be delayed by various political and economic measures it is unlikely that it can be halted unless the existing nuclear powers literally destroy their nuclear arsenals and nuclear weapons establishments. Needless to say such a development is not very likely to take place after the tremendous investments that have been made by several countries in nuclear weapons and nuclear power development.

At present China, France, India, United Kingdom, United States, and Russia have demonstrated their ability to engineer nuclear explosions. Other countries could work toward the achievement of nuclear power status. Dependent on imports of energy from abroad, nuclear power promises an immediate and important source of energy.

At present the United States is the largest uranium producer in the world, with a capacity to produce about 35 percent of the global uranium output. Russia is believed to be the second largest uranium-producing country. Canada, South Africa, and France are the next three largest uranium-producing countries. Niger, Namibia, Australia, and Gabon are the next significant uranium producing countries.

Nonmetallic Raw Materials
Nonmetallic minerals other than the energy minerals such as coal and oil account for only between 5 to 6 percent of the total value among the top 50 most valuable minerals produced in the world. The most important of those are salt, potash, diamonds, phosphates, asbestos, sulfur, kaolin, fluorspar, pyrite, tale, boron, limestone, barytes, mica, feldspar, nitrates, soda, graphite, and asphalt. Diamonds, already discussed earlier in this chapter, are the best known as having specific strategic value, but fertilizer materials that include potash, phosphates, and nitrates are also of crucial importance in the production of fertilizers in countries concerned about increasing the productivity of their agriculture Limestone and fluorspar are important to the steel making process; asbestos is vital in high-temperature thermal insulation, in shipbuilding, and for electrical insulation; tale finds uses in electronics, precision insulators and ultrahigh frequency transmitters.

Depletion of Mineral and Petroleum Resources
Great amounts of iron and copper are now being mined. Is it possible to have large increases of such mining? The highest-grade ore deposits have already been mined, and most high-grade deposits have probably been already discovered. The entire mineral production of the world is increasing exponentially. We are, with improved technology, able to mine lower- and lower- grade deposits. However, as the grade lowers, the cost of production generally goes up. And, this requires a greater use of energy resources. There are limits to what is feasible. Only iron, out of all the metals, is present on earth in good supply; our known iron reserves are enough to last several hundred years. But, obtaining a sufficient supply of nearly all other metals will pose a problem in the future, depending upon the metal in question.

How can we prevent mineral shortages? Is recycling of scrap metals an answer? Yes, but only a partial answer. Substitutions of one material for another may help with some commodities, but generally will probably be only a minor factor. Will more mining of low-grade deposits (the average igneous rock consists of 8 percent aluminum and 5 percent iron), utilizing improved technology, be the answer to our problems? The answer is that it is unlikely, because of the great amounts of energy which must be expended to utilize lower-grade deposits.

We must advance knowledge of the ways that human activities affect natural systems and the environment, and the ways that changing environmental conditions affect humans and the many activities in which they engage. Just as humans influence environmental change through their activities, individuals and organizations can influence environmental change by making decisions that would enable them to reduce adverse impacts on the environment and to adapt more successfully to changing environmental conditions. We must develop tools for assessing policies and options for dealing with environmental change.

A new emphasis will be placed on the examination of policies in innovative modeling frameworks to identify options that are available for responding to environmental change, and to analyze the relative strengths and weaknesses of those options. Results need to be better communicated in a way that contributes directly to the formulation of domestic policy and to the development of international protocols and conventions. Enhanced research on assessing policies and options, therefore, will involve leading social, economic, and policy scientists, whose expertise is necessary to understand the dynamics and interactions of human activities with environmental change, and whose collaboration with natural scientists and engineers is required to develop effectively integrated tools and models. Such collaboration is essential for addressing issues such as predicting the diffusion of new energy-conserving technologies, refining feedbacks within general circulation models that link human-generated emissions with climate, and identifying management strategies that can help preserve natural resources and biodiversity.

It is important to address are the processes through which people become aware of changes in environmental conditions and the ways that they change their behavior in response to those changes through voluntary and coordinated actions. While some effort will be focused on the dynamics of individual decision making, more emphasis will be given to exploring the ways that responses to environmental change are channeled through economic, political, and legal systems as well as through informal organizational structures.

Better knowledge of human-environmental interactions over longer time periods is needed. Many economic models, for example, rely on information gathered by observing the behavior of individuals or groups under specific circumstances. These perspectives are valuable, but they need to be augmented with understanding derived from longer-term anthropological, geographic, and sociological perspectives in order to more fully appreciate the implications of environmental change on people during the coming decades. Similarly, the changing nature of political institutions and protocols for developing international agreements may differ significantly in the future.

Policy science research will examine the ways through which policy questions are framed, including the procedures through which the values of different resources and conditions are established. Different disciplines employ different methods for assigning values, and comparison of these approaches is necessary to determine how the concerns of different stakeholders may be reconciled. Attention also will be given to procedures through which different governments or other kinds of organizations work together to achieve common objectives. Through such research, policy science can provide insights into new approaches that may achieve a more lasting consensus.

Policy science research also will focus on the data, analytic, modeling, and computational needs of policy analysts. The appropriate elicitation and utilization of expert judgments to help fill crucial information gaps will be evaluated. Analytic approaches to be explored include the extraction of general information from case studies, identification of mechanisms for estimating parameters in models, and evaluation of the degree to which assumptions have been maintained in the development and validation of models.

In recognition of the fact that many responses to environmental change must be made at regional to national levels, many of these tools need to be local in character. Because of their sharply defined focus, these tools can be developed, tested, and disseminated quickly. As a result, many new decision tools can be directly employed to address specific problems while subsequent research assesses the degree to which they can be adapted for broader purposes. These new tools will enable national and international decision makers to address competing objectives and to analyze potential trade-offs among available options. There is particularly high demand for new decision tools that can be used in coastal zone management. Management of fisheries, flooding, erosion, water quality and allocation, and sea-level rise in coastal settings requires the development of adaptive strategies.

Recent successes in the refinement of decision-support tools have been possible through advances in knowledge of critical natural processes. As understanding of the frequency and scale of physical phenomena has increased, decision tools have been refined to reduce uncertainties about the socioeconomic consequences of environmental changes. One example is the improved predictive accuracy. Using more accurate forecasts, resource managers have been able to alter management strategies for water use and agricultural practices to reduce economic impacts.

Another example of the value of improved decision tools is the incorporation of site-specific scientific information into economic models. Using new methods, regional distributions of environmental risks can be mapped more accurately, thereby assisting in urban planning and selecting locations for new transportation corridors and industrial facilities. Probability models are examples of successful decision tools which can, as an example, estimate the human impacts of severe rainstorms that reduce the stability of hillslopes. These models demonstrate how site-specific data applied in a regional context can assist emergency-response managers in prioritizing locations for evacuations when rainstorms become severe.

Research Framework

The development of a predictive understanding of how human activities are affecting and affected by the Earth system is among the most complex of all scientific undertakings. This complexity arises for many reasons. Human influences and interactions with the environment cover spatial scales from local to global and extend over time scales from days and seasons to decades and centuries, both back into the past and forward into the future. The influences and interactions involve all sectors of global society and all components of the Earth system, meaning that predictive capabilities must be developed for social and institutional systems as well as physical, chemical, and biological systems. Because the environment is so intimately tied to people's lives, all are impacted by any changes that might occur and all are interested in the character and quality of the predictions. What can be done?

> 1. **Establish** an integrated, comprehensive, long-term program of land, ocean, in situ, and satellite-based observations scale that monitor and describe the current state of the environmental and analyze data on recent and past actions on the environment;

> 2. **Manage** and archive information, which includes assembling, processing, storing, and distributing data and information that documents the state of the environment, both the conditions of the natural system and of the societal systems that are influencing and are influenced by environmental changes;

3. **Understand** Processes, which includes conducting a program of focused studies to measure, analyze, and investigate the physical, chemical, biological, geological processes and societal influences that govern behavior and the interactions of human activities with the global environment;

4. **Predict** Changes, which includes developing, testing, and applying integrated conceptual and predictive models of the environmental system in order to provide insights and projections of the response of the atmosphere, oceans, and land surface to natural and human influences and to reconcile predicted and observed Earth-system behavior;

5. **Analyze** Consequences, which includes evaluating and interpreting the environmental and societal consequences and impacts of change and understanding the potential for, and limitation to, natural and technology-enhanced adaptation to and mitigation of environmental change; and

6. **Assess** Policies and Options, which includes research on social and economic interactions, decision frameworks (especially those which include decision making under uncertainty), and development of the policy and economic tools of analysis to examine the relative strengths and weaknesses of the various choices for responding to environmental change.

These streams of research activity can provide scientific information in support of national and international decision making concerning the broad spectrum of natural and human-induced changes occurring in the global and regional environment.

To support coupling of engineering activities to the national and international community, there are two activities focusing on outreach and integration:

1. **International Interactions**, which includes encouraging and promoting cooperation with other nations in developing

scientific understanding and establishing the institutional framework necessary for broad-based consideration of environmental change issues;

2. **Education and Public Awareness**, which includes preparing materials and organizing activities that promote consideration of environmental change and its human dimensions as part of both public awareness and educational processes.

Because of the need for improving understanding of the complex and diverse relationships between human activities and environmental change, we must support programs that develop tools for assessing policies and options, especially in the area of integrated assessment methods. It also will increase support for research in the social, economic, and policy sciences, and for research that examines the impacts of and responses to environmental change. With higher levels of support, we will be able to improve understanding of the impacts of environmental change on human health, economic systems, settlements, and societal structures. Increased support in programs will also facilitate research on options for mitigation and adaptation strategies and the development of technologies to implement those strategies. Most significantly, new research will result in better connection of research to policy making. Of special interest to policy makers will be the development of new decision analysis tools and methodologies for integrating assessments of environmental change, its impacts, and potential response options.

Linking research to policy requires the development of new programs on the human dimensions of environmental change. New social and economic research will focus on present and future patterns of greenhouse gas emissions, actions that can be taken to modify those patterns, and the effectiveness and potential benefits and costs of these actions. Contributing research will improve understanding of the effects on natural systems of factors like population growth, economic growth, technological change, and international trade.

Just as uncertainties remain with respect to the response of the climate system to changing emissions of greenhouse gases, important gaps in knowledge persist as to how people and institutions respond to changes in natural and human environmental conditions. These uncertainties will affect future levels of emissions, and they will influence the consideration and effectiveness of specific policies to reduce those emissions.

Research in the natural sciences can provide only limited information for understanding the causes and impacts of, and societal responses to environmental earth engineering. Research in the natural sciences needs to be integrated with research in the social, economic, and behavioral sciences, the policy sciences, and the health sciences in order to provide a more complete understanding of the human dimensions of environmental change.

Research on earth engineering is concerned with activities that drive environmental change and on activities that will be affected by environmental change. A number of specific questions confront policy and decision makers in governments and in many different economic sectors as they consider the management and use of resources such as energy, land, water, soils, and forests. Foremost among these questions are "How is environmental change likely to affect the availability and quality of resources?" and "How will policies and actions designed to mitigate or adapt to environmental change affect the supply, cost, and use of resources?" Research on resource management is especially important in situations where resources have multiple uses. The vulnerability of large-scale resource management investments to environmental change, especially in developing countries where resource investments often provide critical and life-sustaining services, is another important issue.

Another priority is the valuation of environmental goods, services, and assets, which requires fundamental research on how changes in the availability and quality of these resources are understood and measured in social terms. Environmental accounting possesses significant definitional and characterization problems. Unlike economic systems, environmental systems do not have a calibrating mechanism such as market prices to determine relative unit values. While numerical values exist for environmental goods, the consistency and interpretation of these values is a problem. Existing techniques to measure environmental values, such as contingent valuation, will be the focus of further development, testing, and validation. Alternative paradigms of valuation, derived from other social and behavioral sciences that reflect the psychological, cultural, and ethical dimensions of environmental values, will be studied.

Policy science research must focus on data, analytical methods, computational issues, and modeling requirements of policy analysis. Significant data issues include availability, organization, validation, reliability, and use of expert judgment techniques to fill data gaps. Analytical issues include developing methods for extracting general relationships from case-studies, which are important for historical evidence as well as for studies on environmental services and benefits.

Chapter 3. Global Climate Change

"Look deep into nature, and then you will understand everything better."—Albert Einstein

Global Change

Human activities have long influenced local environments where people live. Since the start of the Industrial Revolution and the attendant rapid population explosion, human activities also began to influence the global environment. Human perception of climate is the result of many local and time-varying experiences, ranging from daily variations in rain and shine to yearly variations in the intensity of the seasons. Climate influences the local nature of clouds and storms, the quality and quantity of water resources, the productivity of agricultural regions, and other activities of societal concern. Agricultural productivity and economic activities are closely tied to the seasonal climate, and can be significantly affected by the extremes of floods, drought, and severe weather, as well as the steady change of average climatic conditions. Although different climates have existed in the past, it is the projected increase in the rate of change and its global extent that make climate change a critical issue. The changes resulting from these activities have significant environmental consequences. Three important problems are

1. Carbon dioxide and global warming

2. Oz Depletion and Ultra Violet (UV) Radiation

3. Variations in the Seasonal Climate

These three problems are by no means the only ones of global environmental concern. Another is the extension of agriculture and rapid increases in population that are leading to major changes in land use, including deforestation and dryland degradation, which often have detrimental effects on the resilience and complexity of ecosystems. Another is the development of coastlines which is altering beach processes, reducing coastal habitats, and making

Chapter 3. Global Climate Change

communities more vulnerable to severe weather and sea level change.

Carbon Dioxide and Global Warming

Oxygen is continuously replenished in Earth's atmosphere by photosynthesis, which uses the energy of sunlight to produce oxygen from water and carbon dioxide. Oxygen is too chemically reactive to remain a free element in air without being continuously replenished by water and carbon dioxide in the air.

Photosynthesis is a process used by plants and other organisms to convert light energy into chemical energy that can later be released to fuel the organisms' activities. This chemical energy is stored in carbohydrate molecules, such as sugars, which are synthesized from carbon dioxide and water – hence the name photosynthesis. The word comes from the Greek, phōs, "light", and synthesis, "putting together." In most cases, oxygen is also released as a waste product. Most plants, most algae, and cyanobacteria perform photosynthesis; such organisms are called photoautotrophs. Photosynthesis is largely responsible for producing and maintaining the oxygen content of the Earth's atmosphere, and supplies all of the organic compounds and most of the energy necessary for life on Earth. If carbon dioxide were to disappear, all life on earth would die (except for lifeforms that do not depend upon oxygen).

An organism that doesn't require oxygen to survive is known as an anaerobe. There are two kinds of anaerobes: facultative anaerobes and obligate anaerobes.

> 1. Facultative anaerobes can live with or without oxygen. In the presence of oxygen, they make energy using aerobic cellular respiration. In the absence of oxygen, they switch to fermentation. Staphylococcus and E.coli are facultative anaerobic bacteria, and yeast is a facultative anaerobic fungus. There are also aquatic worms that are facultative anaerobes.

2. Obligate aerobes cannot grow in the presence of oxygen, and many of them will die if exposed to O2. Clostridium botulinum, the bacterium that causes botulism food poisoning and produces the active ingredient in Botox treatments, is an obligate anaerobe.

Since the industrial revolution, the carbon dioxide content of the atmosphere has been steadily increasing. This increase in the global atmospheric carbon dioxide concentration is due to human activities such as the burning of fossil fuels and changes in land-use practices. People have increased the carbon dioxide content of the atmosphere more than 10 percent in the last century.   Carbon dioxide plays an important role in controlling the Earth's temperature and weather. An increase in carbon dioxide may lead to a significant increase in the average temperature of the Earth through the so-called greenhouse effect. The greenhouse effect is the popular term referring to the trapping of infrared (heat) energy in the atmosphere.

Let us explain how the greenhouse effect works. We can see through a transparent object. However it is not necessary that all kinds of light can pass through a transparent object. For example, red glass is considered transparent because we can see through it, but red glass will not transmit blue light. Ordinary glass, which is transparent to all colors of light, is only slightly transparent to ultraviolet or infrared light. Visible sunlight passes through a glass house and is absorbed by the material within the house. As a result, the material is heated. In return, the material warmed by the sunlight radiates that warmth. However this radiation is not in the form of visible light, but in the form of the much less energetic infrared radiation. Only small quantities of the infrared radiation go through the glass. Most is reflected by the glass back onto the material, so that energy accumulates within the glass house. Once equilibrium is reached, the material radiates as much infrared energy as it absorbs in the form of sunlight, and its temperature remains constant.   This temperature is higher that it would be if the sun was not shinning on the material.   If the material were in the open, it would easily get rid of its infrared radiation, but the sun-warmed material within the glass house experienced continual reflection. As a result the temperature of the material within the house rises much higher than does the temperature of material outside. In fact, the temperature inside rises until enough infrared radiation can leak through the glass to set up an equilibrium point. Because of this, plants can be grown inside a glass house in the winter. For this reason a glass house is called a greenhouse. Thus we can describe the greenhouse effect as the additional warmth inside the greenhouse caused by the fact that glass is transparent to visible light and only slightly transparent to infrared radiation. The carbon dioxide of the atmosphere reflects solar energy that is reradiated from the ground, and in turn part of this heat is reradiated back to Earth.

The atmosphere of the Earth consists almost entirely of oxygen, nitrogen, and argon. These gases are transparent to both visible light as well as to the infrared radiation that the Earth's surface gives off when it is warmed. The atmosphere also contains 0.03 percent of carbon dioxide. The carbon dioxide that is present in our atmosphere makes the Earth warmer than it would be if there were no carbon dioxide present at all. If the carbon dioxide content of the atmosphere were to double, the increased greenhouse effect would warm the Earth a couple of additional degrees.

Given the size and tremendous heat capacity of the global oceans, it takes a massive amount of accumulated heat energy to raise Earth's average yearly surface temperature even a small amount. Behind the seemingly small increase in global average surface temperature over the past century is a significant increase in accumulated heat. That extra heat is driving regional and seasonal temperature extremes, reducing snow cover and sea ice, intensifying heavy rainfall, and changing habitat ranges for plants and animals—expanding some and shrinking others.   In the absence of offsetting factors such as variable solar radiative output, climate models predict that increases in greenhouse gas concentrations will raise the temperature of the Earth with potentially disruptive consequences. Reconstructions of past changes in climate over the Earth's history and theoretical understanding of how atmospheric, oceanic, and biological processes combine to determine the climate suggest that the increase in global average temperature over the next century.   Also zones of precipitation may shift, generally poleward and perhaps closer to coastal regions.

Though warming has not been uniform across the planet, the upward trend in the globally averaged temperature shows that more areas are warming than cooling. According to the NOAA, the combined land and ocean temperature has increased at an average rate of 0.07°C (0.13°F) per decade since 1880; however, the average rate of increase since 1981 (0.17°C / 0.31°F) is more than twice as great.

We thus come to the idea of too much carbon dioxide in the air. The consequence would be that the Earth's climate would undergo a warming because of the greenhouse effect. Increases in the concentrations of carbon dioxide and other gases are enhancing the Earth's greenhouse effect. The addition of heat in this manner could cause the world's present glaciers to melt, thereby raising sea level and drowning the seaports and coastlines of the world. Heating the Earth's atmosphere by a small amount could result in an unfrozen Arctic Ocean, hence more evaporation and precipitation (snow) in the northern polar region with a resulting snow accumulation and glacial advance.

Let us now consider impacts on climate change. The changes in concentrations of greenhouse gases are predicted to induce a wide range of climatic changes in addition to global warming. Warming of the oceans and the melting of icecaps and glaciers will result in sea level rise. The amount of sea level rise over the next century is projected to be tens of centimeters (several times the rate of rise in the recent past), at the very least increasing the threats to coastal areas from storms and hurricanes. Prospective shifts in precipitation patterns are predicted by some models to lead to important shifts in world agricultural regions and in the availability of water resources, with the possibility of altering long-established patterns of land use. At the same time, the growth rate of plants under some conditions can be increased in the presence of additional carbon dioxide, and

Chapter 3. Global Climate Change

forests and grasslands can also be affected by increased carbon dioxide and shifted precipitation patterns. Together, these changes have the potential to cause important shifts in flora and fauna. It is vital to understand what the vulnerabilities and capabilities of these systems to adjust are, and how their resilience to change can be enhanced.

Because projected acceleration in the rate of change is vitally important in the greenhouse effect, it is necessary to monitor and understand the changes in concentrations of radiatively active gases and aerosols in the Earth's atmosphere and to quantify the effects of those changes on the radiative forcing of climate. In addition to carbon dioxide, other gases act as greenhouse gases, the major ones being CH_4, N_2O, the CFCs and their substitutes, and ozone in the lower stratosphere and troposphere. The atmospheric concentrations of all of these gases are changing as a result of human activities. Recent findings have provided important new insights:

Carbon Dioxide. The atmospheric concentration of carbon dioxide (CO_2) has risen about 30% since the 1700s. This increase is responsible for more than half of the enhancement of the trapping of infrared radiation due to human activities. Over the past two years the rate of rise in the carbon dioxide concentration has, surprisingly and inexplicably, slowed. There is reason to believe that this reduced rate of increase in carbon dioxide concentration may be short-term. New measurements are refining estimates of ocean uptake, but important uncertainties remain about uptake of carbon dioxide by terrestrial vegetation and soils.

CFCs. Because CFCs are believed to destroy lower stratospheric ozone, which is also a greenhouse gas, the net effect of CFC's as greenhouse gases is less than previously believed. Observations show the rate of increase of CFCs in the atmosphere to be slowing, consistent with international emissions controls, although stratospheric ozone depletion continues to occur.

Methane. The atmospheric lifetime of methane (CH_4) has been determined to be 25% longer than previously thought, which contributes to raising its global warming potential (GWP). New research has shown that while wetlands are a significant reservoir for carbon, methane emissions from wetlands may increase with increasing carbon dioxide concentration in the atmosphere. Over the last few years, the rate of increase in atmospheric methane has slowed. In 1992, the rate of increase was sharply reduced, but current measurements indicate that it is returning to earlier values. Continued research is needed to understand whether the reduced rate was due to reduced human emissions or to an enhancement of sinks.

Ozone. Tropospheric ozone is an important greenhouse gas, and its influence is greatest in the upper troposphere. Calculations suggest that increases in tropospheric ozone can substantially increase radiative forcing. However, trends in tropospheric ozone still have considerable uncertainty. Field measurements demonstrate that most tropospheric ozone over the temperate North Atlantic Ocean is derived from transported North American presursors (nonmethane hydrocarbons and reactive nitrogen compounds).

Very long-lived greenhouse gases. The lifetimes of fully-fluorinated (perfluorinated) hydrocarbons (PFCs), such as CF_4, C_2F_6, C_6Fl_4, have been shown from laboratory and modeling studies to exceed a thousand years. Understanding their chemistry and lifetimes are critical because some have been proposed as CFC substitutes, while others are emitted as trace products of industrial processes, including aluminum production.

Aerosols. The atmospheric burden of aerosols from human activities continue to be quite high. Anthropogenic aerosols include sulfate aerosols, which are formed primarily from sulfur dioxide emitted from major urban and industrial complexes, and carbon-containing aerosols, which are emitted from combustion of fossil fuels and from biomass burning. The impacts of an increasing atmospheric aerosol burden, on a regional scale, include reduced solar radiation at the surface (which can lead to cooling), and changes in atmospheric dynamics (which can lead to variations in seasonal rainfall and temperature patterns). Modeling studies suggest that, in contrast

Chapter 3. Global Climate Change

to greenhouse gases, anthropogenic sulfate aerosols can lower surface temperatures. Research on the radiative effects of atmospheric aerosols resulting from emissions from coal and oil combustion and heavy industrial processes is important to understanding whether aerosols may be, in the near term, counterbalancing the enhanced greenhouse effect of carbon dioxide. The absence of measurements confirming the predicted increase in land surface temperatures in the Northern Hemisphere appears to be most related to recent increases in the frequency of cloud cover. Recent studies suggest that the hemispheric asymmetry in this century's warming may be due, at least in part, to the preferential presence of sulfate aerosols in the Northern Hemisphere as a result of industrial emissions patterns. For carbonaceous aerosols emitted by biomass burning, the sign of their climatic effect is less certain as is predicted by the research results. The atmospheric burden of natural aerosols was reduced in 1993. Both space- and ground-based measurements showed that the volcanic aerosols injected into the stratosphere by the June 1991 eruption of Mt. Pinatubo in the Philippines are finally being removed from the atmosphere by natural processes. This has resulted in the return of the global average temperature to the warmer levels typical of the 1980's, and to a reduction in the sharp decrease in ozone concentration in the lower stratosphere that is thought to have been caused by volcanic aerosols. Research is ongoing to establish a baseline of information about naturally forming aerosols in remote maritime locations to which the climatology of aerosols in more populated areas can be compared. The program, a cooperative effort between government laboratories and universities, includes ship cruises to gather information about aerosol formation in ocean areas of both hemispheres, and the establishment of several monitoring stations in both continental and maritime areas to provide long time series of aerosol data.

As we have seen, greenhouse warming relates to the potential for greenhouse gases and aerosols emitted as a result of human activities to alter the global climate and cause significant impacts on the natural environment and societal activities. However greenhouse warming may be mitigated by other effects. Evidence has shown that increases in atmospheric aerosols, which can scatter solar energy and alter cloud cover, may offset some of the effects of increases in greenhouse gas concentrations. Because of the inhomogeneous distribution of aerosols and because of uncertainties in their influence on cloud formation and hence on radiative transfer, their impact on climate remains difficult to quantify. Quantitative understanding of the sources and sinks of atmospheric trace gases and aerosols and of the interactions which contribute to changes in biogeochemical processes are crucial for making reliable predictions of future concentrations of atmospheric greenhouse gases.

Stratospheric Ozone Depletion and UV Radiation

Stratospheric Ozone Depletion and UV Radiation relates to the effects of emissions from human activities on the atmospheric ozone layer, and the consequent reduction in the ability of the atmosphere to screen out ultraviolet (UV) radiation.

Ozone.
Ozone is a gas made up of three oxygen atoms. It occurs naturally in the stratosphere where it protects life on Earth by absorbing the sun's ultraviolet radiation. Ozone measurements indicate that tropospheric ozone derived from anthropogenic pollution may exceed that derived from natural sources over the North Atlantic. Long-range transport of continentally-derived precursors can account for as much ozone on a hemispheric scale as do natural processes. Because upper tropospheric ozone is an extremely potent greenhouse gas, we must consider global warming in the context of regional pollution abatement. We must understand the role of biomass burning and stratospheric inclusions in determining the high ozone concentrations occurring over the South Atlantic. It is necessary to study boundary layer fluxes of ozone and exchanges with the free troposphere both in the tropics and at high northern latitudes.

Ozone (trioxygen) is an inorganic molecule with the chemical formula O_3.. It is a pale blue gas with a distinctively pungent smell. It is an allotrope of oxygen that is much less stable than the diatomic

Chapter 3. Global Climate Change

allotrope O_2, breaking down in the lower atmosphere to O_2 (dioxygen). Ozone is formed from dioxygen by the action of ultraviolet light (UV) and electrical discharges within the Earth's atmosphere. It is present in very low concentrations throughout the latter, with its highest concentration high in the ozone layer of the stratosphere, which absorbs most of the Sun's ultraviolet (UV) radiation.

Ozone's odor is reminiscent of chlorine, and detectable by many people at concentrations of as little as 0.1 ppm (i.e.,. parts per million) in air. In standard conditions, ozone is a pale blue gas that condenses at progressively cryogenic temperatures to a dark blue liquid and finally a violet-black solid. Ozone's instability with regard to more common dioxygen is such that both concentrated gas and liquid ozone may decompose explosively at elevated temperatures or fast warming to the boiling point. It is therefore used commercially only in low concentrations.

Ozone is a powerful oxidant (far more so than dioxygen) and has many industrial and consumer applications related to oxidation. This same high oxidizing potential, however, causes ozone to damage mucous and respiratory tissues in animals, and also tissues in plants, above concentrations of about 0.1 ppm. While this makes ozone a potent respiratory hazard and pollutant near ground level, a higher concentration in the ozone layer (from two to eight ppm) is beneficial, preventing damaging UV light from reaching the Earth's

Let us now explain the ozone hole. The Antarctic ozone hole was discovered in 1985. The Antarctic ozone hole is an area of the Antarctic stratosphere in which the recent ozone levels have dropped to as low as 33 percent of their pre-1975 values. The ozone hole occurs during the Antarctic spring, from September to early December, as strong westerly winds start to circulate around the continent and create an atmospheric container. Within this polar vortex, over 50 percent of the lower stratospheric ozone is destroyed during the Antarctic spring.

The atmospheric ozone layer is located in the stratosphere at an altitude of 16-18 km at the poles and up to 25 km near the equator. It has an important role in adsorbing ultraviolet (uv) radiation. Recent evidence suggests that this layer of ozone is being depleted by man-made chemicals, especially the chlorofluorocarbons (CFCs). CFCs are widely used as propellants, solvents, and as cooling agents in refrigerators and air conditioning systems. The bromine-containing halons are also suspected of causing ozone depletion. A decreased protection from ultraviolet radiation due to reduced concentrations of ozone in the stratosphere is expected to lead to an increased incidence of skin cancer.

The Antarctic ozone hole grows and shrinks with the seasons. It begins to open in August. By October, the hole reaches its maximum size, and it closes by December. Ice clouds that form in the stratosphere during the Southern Hemisphere's winter break down the harmful chemicals into chlorine and bromine. When spring arrives, sunlight activates the chlorine and bromine, which causes rapid loss of ozone. The Montreal Protocol was signed in 1987 to ban the use of the main human-made ozone-depleting chemicals. Nations agreed to phase out the production of substances that lead to ozone destruction. The Montreal Protocol, which was ratified by 196 countries and the European Union, monitors the production of these chemicals that were commonly found in aerosol sprays, refrigerators and air conditioning units. Consequently, concentrations of these substances are declining, and it is expected that Antarctic ozone levels to return to pre-1980 values by 2060.

Significant increases in ultraviolet (UV) radiation have been observed in conjunction with periods of intense ozone depletion. The ozone hole over Antarctica, which in 1993 produced the lowest values of ozone ever recorded on Earth, also allowed record levels of UV light to reach Antarctica. At one Antarctic monitoring site, UV-B, the part of the spectrum most harmful to life, has been recorded at levels 44 percent higher than in 1992. Investigations are now underway on the impact that the increased UV might have on life on and around Antarctica, and on whether animals and plants may have

Chapter 3. Global Climate Change

mechanisms to avoid harm from increased UV. Studies have already estimated that UV damage has reduced the productivity of ocean phytoplankton— tiny plants that comprise the base of the food chain—by 6% to 12% in areas affected by the ozone hole.

A worldwide network of Dobson spectrometers monitor atmospheric ozone concentrations. Ozonesondes are also used to determine the vertical distribution of ozone in the atmosphere. Stratospheric networks are designed for observing and understanding the physical and chemical state of the stratosphere, with special emphasis on the depletion of stratospheric ozone.

Ozone-layer depletion, the associated increase in ground-level ultraviolet radiation, and the impacts on human health and biota are significant environmental problems. Human-produced chemicals containing chlorine and bromine, collectively referred to as halocarbons, are depleting the stratospheric ozone layer. Even if control measures are fully implemented, ozone depletion will continue for nearly another decade. Because of the long atmospheric lifetimes (up to 100 years) of many of the halocarbons, the earliest recovery from the Antarctic ozone hole is several decades away, and a return to near the natural atmospheric levels of chlorine and bromine, and therefore of ozone, will take centuries. Ozone-layer depletion is also known to be linked to global climate change.

Significant research is being conducted to provide information about the health effects on humans of increased UV radiation and CFC substitutes. Health studies on the effects of increased UV exposure focus on the impacts on the immune system, aging process, sensitive tissues, and methods to reduce these harmful effects. The principal objectives of these studies are to promote an increased understanding of UV effects on target organs (e.g., eyes and skin) and the molecular changes that lead to these effects, and to help develop strategies to prevent the initiation of disease or to intervene before disease.

Animal models indicate that exposure to environmentally relevant doses of UV-B radiation can adversely affect the course of certain infectious diseases. Exposure to UV-B can drastically reduce survival time after exposure to lethal agents. Other research has shown that the ability of UV radiation to impair the development of cell-mediated immunity depends on the particular antigen administered, and that DNA is the primary target of UV radiation in the generation of systemic immunosuppression.

Long-term global satellite and ground-based monitoring activities have demonstrated that stratospheric ozone depletion is occurring over most of the globe, except in the tropics. Downward trends of several percent per decade are now observed in all seasons at mid-latitudes (poleward of 20°), with winter and springtime declines of as much as 6 - 8% per decade observed poleward of 45°.

Atmospheric concentrations of greenhouse gases and ozone depleting substances are monitored through an informal worldwide network of in situ and flask-sampling sites. Long-term measurements, from which global trends are inferred, include carbon dioxide, methane, nitrous oxide, and halocarbons. Significant recent findings include the following:

>1. Measurements at globally distributed observatories have revealed a sharp increase in atmospheric concentrations of HCFC22, a major CFC substitute, demonstrating that these substances are beginning to accumulate in the atmosphere. A marked inter-hemispheric gradient has also been observed. These data not only provide an important check on the human use and atmospheric fate of this compound as called for by the Montreal Protocol, but also yield new information on transport between the two hemispheres;

>2. Thousands of globally distributed measurements have shown that the rate of increase of atmospheric methane slowed down substantially, suggestive, but not conclusive of, changes in human emissions; and

>3. Observations in 1993 confirmed a significant reduction in the rate of increase in the atmospheric carbon dioxide

concentration. Annual increases in the mean concentration of carbon dioxide were smaller in the years 1991 and 1992 than in any other year since continuous monitoring was begun in 1957. By 1993, the annual rate was about 25% of what it had been over the past decade. While carbon dioxide accumulation varies from year-to-year, the magnitude and persistence of this reduction are unprecedented in the modern record, and is thought to be due to increased carbon storage in the terrestrial biosphere. Whether these changes will persist is an important unanswered question.

The Antarctic Situation at September 2019 is described as: The annual stratospheric warming has commenced unusually early this year and temperatures are rising through much of the ozone layer. The area above Antarctica with Polar Stratospheric Clouds (PSCs) has dropped to only 1 million square kilometers, the smallest at this time of year for decades. The polar vortex has begun to shrink and is now 24 million square kilometers in area near the base of the ozone layer; this is smaller than it has been in the last decade. It is shrinking more rapidly higher in the ozone layer. The vortex is offset from the Pole towards the Atlantic. The lowest ozone amounts are over West Antarctica. Ozone amounts are much higher around Antarctica over the southern ocean.   The ozone hole is currently some 5 million square kilometers in area, down from a peak of around 11 million square kilometers and smaller than ever seen in the last decade.

Carbon Dioxide
The annual increase in atmospheric concentrations of carbon dioxide is the difference between the emissions of carbon dioxide from combustion of fossil fuels and biomass and uptake by oceans and the biosphere. It is believed that enhanced sinks for carbon are responsible for the recent slowdown in the carbon dioxide increase in the atmosphere, but the exact cause remains unexplained. Recent evidence suggests that in the short term, at least, the land sink may be larger than previously thought. Anomalies in surface temperature and precipitation may have allowed terrestrial ecosystems to accumulate more carbon in the period 1991 to 1993

than normally would have been the case. The cause of the anomaly is unclear, but the timing is coincident with the eruption of Mt. Pinatubo. One hypothesis is that lower temperatures at the Earth's surface temporarily reduced rates of respiration and thereby enhanced net carbon uptake.   Because few measurements of the carbon concentration in coastal oceans have been made, there is only limited understanding of the role of continental shelves in the exchange of carbon dioxide between the coastal shelf and the atmosphere above it, even though half of ocean photosynthesis is estimated to occur on continental shelves.   The continental shelves could be a relatively large sink for atmospheric carbon dioxide during certain times of the year.   It is necessary to obtain a high accuracy snapshot of the distribution of carbon dioxide within the world's oceans.   This data can be used to calculate the exchange of carbon dioxide between the ocean and the atmosphere. Together with models of the atmospheric transport of carbon dioxide, and with an improved understanding of the carbon dioxide fertilization effect, these calculations will allow the locations and magnitudes of net sources or sinks of carbon dioxide to be inferred on a planetary scale. This approach will also permit improved estimation of the uptake of fossil-fuel carbon dioxide by the oceans, both through modeling and through correlation with the distribution of other man-made compounds within the ocean.

Perfluorinated Hydrocarbons
Laboratory measurements have demonstrated that perfluorinated hydrocarbons (PFCs), which are greenhouse gases emitted as byproducts of industrial processes such as aluminum production and proposed as CFC substitutes, persist in the atmosphere for thousands of years. As a result, their emissions into the atmosphere are essentially irreversible, and their contributions to global warming are inadequately characterized by short-time global warming potentials. Proper consideration of the environmental consequences of PFC emissions, therefore, imposes a new requirement to grapple with extremely long-term effects. Refined methods for calculating ozone depletion potentials must be

Chapter 3. Global Climate Change

developed using atmospheric measurements rather than only model results.

Aerosols
Emissions into the atmosphere of aerosols, and gases that chemically react to form aerosols, can have direct effects both on the global radiation balance and on global atmospheric chemistry. Research on the physical and chemical characteristics of tropospheric aerosols is underway to determine whether changes in aerosol (and aerosol precursor) emissions may be, at least temporarily, hiding the warming effect of greenhouse gases. Research on the radiative effects of sulfate particles formed in the lower troposphere, mainly as a result of emissions from coal combustion, is important to understanding whether they may be, in the near term, counterbalancing the enhanced greenhouse effect of carbon dioxide. For aerosols emitted by biomass burning, the sign of their climatic effect is less certain, being dependent on the amount of black carbon in the aerosol. The absence of measurements confirming the predicted increase in land surface temperatures in the Northern Hemisphere appears to be most related to recent increases in the frequency of cloud cover. Recent studies suggest that the hemispheric asymmetry in this century's warming may be due, at least in part, to the anthropogenic aerosol emissions being largely in the Northern Hemisphere.

Methane
Atmospheric methane concentrations have been increasing at a reduced rate. The residence time of methane in the atmosphere has been found to be 25% longer than previously thought, however. Research is ongoing to quantify human-induced methane emissions from landfills, coal mines, natural gas systems, rice paddies, and biomass burning as well as natural emissions of methane from wetlands and other sources. The results of this research will provide baseline data to help understand the causes of the increasing concentrations of atmospheric methane and for identifying strategies for reducing emissions from various sources.

Regional measurements of carbon dioxide uptake by forest vegetation suggest the net uptake of carbon worldwide could account for the estimated missing carbon in the global carbon budget. Based on almost two years of continuous measurements of carbon dioxide exchange between the atmosphere and the vegetation and soils of Harvard Forest in eastern Massachusetts, the net annual uptake of carbon by the forest ecosystem was estimated to be as large as 4 tons per hectare. Confirmation of these results with other natural vegetation studies would support the use of forest management (including reforestation) as an interim mitigation strategy that is an environmentally attractive way of helping to reduce the buildup of atmospheric carbon dioxide. Uncertainties regarding the rate of deforestation, conversion of land to agricultural use, population growth, and land management practices make it difficult to estimate the extent to which forests can conserve or sequester large quantities of carbon on a global basis.
International efforts are underway to improve understanding of the exchange of gases, energy and water between the boreal forest biome and the atmosphere in order to clarify their roles in global change. Airborne and space-borne remote sensing can be used to extend understanding of these processes from the local to regional scales.

Recent research has documented that the uptake of carbon by soils is a dynamic process that varies in relationship to soil age and type. Understanding the role of soils in the carbon cycle is critical to understanding how human activities, including agricultural and forestry practices, influence the fluxes of carbon from terrestrial ecosystems. Remotely sensed surface temperature over a 300-km by 300-km region in western Montana has shown that the ratio of surface temperature to a normalized difference vegetation index can distinguish wet and dry conditions in a forest. This ratio may have applications in climate models, research on decomposition, and monitoring of fire conditions.

Recent studies focus on the effects of global change (e.g. climate) on ecosystems, and the feedback effects of ecological processes on

Chapter 3. Global Climate Change

atmospheric composition (e.g., greenhouse gas concentrations) and climate. Its scope includes research on how ecological processes are affected by altered atmospheric carbon dioxide and other trace gases, by altered climate conditions, and by changing land-use patterns, all in relation to constraints of other resources (e.g. nutrients, water, light, etc.). This initiative also includes research on how ecological systems affect the exchange of carbon dioxide and other trace gases with the atmosphere and ultimately determine terrestrial sources and sinks of carbon. The following programs are underway:

- Experiments that determine ecosystem responses to combined effects from global forcing of elevated carbon dioxide, temperature, water, and nutrients;

- Field studies designed to provide a predictive understanding of the combined effects of global forcing (e.g., climate change) and landscape-scale processes on the future structure and distribution of ecosystems;

- Research to determine the potential effects of global forcing on the biodiversity (e.g., species diversity, genetic diversity, habitat diversity) of managed and unmanaged ecosystems, and in turn, what the resulting effects of biodiversity changes on ecosystem function will be.

Volcanic Cooling

Volcanic eruptions can contribute large quantities of gases and aerosols to the Earth's atmosphere and have been linked to past climate change. The USGCRP Global Volcanism Program maintains a database of the world's volcanoes and their known eruptions of the last 10,000 years; an archive of maps, photographs, and other historical documentation of the world's volcanoes; and a Global Volcanism Network, a network tracking and reporting current volcanic activity around the world.

Stratospheric ozone depletion is linked to changes in the surface climate. Loss of lower-stratospheric ozone is predicted to lead to a cooling tendency at the surface. As a result of this effect, ozone decreases offset some of the greenhouse warming of the halocarbons that caused the ozone change. Such indirect couplings complicate projection of changes in the global climate.   Let us explain.

Global ozone depletion was observed to be significantly worse in 1992 and 1993, including wintertime depletions of up to 25% over populated regions in the high-latitudes of the Northern Hemisphere. The observations of unexpected and unprecedented ozone depletion in the past two years, coinciding with the period following the eruption of Mt. Pinatubo, have revealed new gaps in understanding and, hence, in prognostic capabilities. While ozone levels may have been perturbed by the Mt. Pinatubo eruption, either by changes in stratospheric temperature and/or circulation, or by enhanced heterogeneous chemistry, the magnitude and timing of the recent, large ozone decreases are not fully explained by the current understanding of these effects. Consequently, evaluation of the heterogeneous chemistry associated with surface reactions on aerosols through laboratory studies, atmospheric observations, and modeling remains a key research priority that requires an enhanced focus.   The increasing concentrations of greenhouse eases in the atmosphere are not the only factors influencing climate. Explosive volcanic eruptions can inject enormous amounts of sulfur dioxide and ash into the atmosphere. Aerosol particles injected into the stratosphere, can result in climate changes lasting up to several years. Observed climatic responses to the Mt. Pinatubo eruption have included tropospheric cooling, stratospheric warming, and an overall drop of about 0.5 °C in the global average surface temperature.

There are not yet comprehensive estimates of how the effects of changes in aerosol concentrations, changes in land cover and land use, and changes in concentrations of greenhouse gases will combine with natural influences to alter the global climate. Examination of the temperature record of the last 100 years does show a warming of about 0.5 °C, only temporarily reversed recently by the volcanic influence of Mt. Pinatubo, suggesting that the enhanced greenhouse effect is exerting the primary influence. The fact that this warming is somewhat less and different in timing than that predicted by computer models emphasizes the need for continuing research directed toward gaining a better understanding of both human and natural influences such as solar variability on the climate system.

Significant Variations of the Seasonal Climate

Significant variations of the seasonal climate refers to the agricultural, economic, and related effects on human activities of sharp fluctuations and variations in the seasonal to inter-annual climate, particularly the extended heavy precipitation and drought episodes associated with El Nino-Southern Oscillation (ENSO) events in the tropical Pacific Ocean. In 1982-83, the world experienced the largest, most severe El Nino of this century. No one predicted it. In fact scientists did not even realize it was underway until several months after it started Anomalous weather events all over the world in 1982-83 were attributed to this El Nino. In early 1986, oceanographers issued the first scientific forecasts for a moderate El Nino by late summer 1986. El Nino occurred nearly as predicted.

What Is El Nino? In Ecuador prior to 1960, anglers looked forward to a warming of the nearshore waters, which would bring some different fish to their nets for Christmas. (Hence the name, El Nino, Spanish for the child.) After the 1957-58 large climate changes in the tropical Pacific Ocean and atmosphere, U.S. scientists adopted the term El Nino to designate the occurrence of a large body of warm water in the Pacific off the South American coast. This phenomenon occurs every three to seven years, with no regular periodic cycle. When a moderate to strong El Nino occurs, the ocean off Peru and Ecuador for thousands of square miles becomes several degrees warmer than normal. The usual strong atmospheric convection over Asia and Indonesia shifts to the central and eastern Pacific. All types of unusual climate features ensue. For example, many global flood and drought areas are affected by El Nino. People all over the world would benefit from accurate forecasts of El Nino. For example, it might be wise to limit anchovy fishing in Peru because El Nino will kill much of the new generation of fish and drive away the adult fish usually caught. The El Nino could help to forecast a milder and wetter winter in the southeastern United States. We could predict a decrease in the number of hurricanes in the Atlantic if an El Nino is forecast. There could be a decreased coconut oil supply in Indonesia. In Australia there could be a high likelihood of severe drought. In Brazil we might forecast both unusual floods and drought.   In effect, El Nino is blamed for almost all unusual climatic occurrences on the planet.

Efforts to model El Nino are based on the hypothesis that the tropical Pacific Ocean responds robustly to the large trade wind systems. In the Northern Hemisphere, the northeast trade winds cover most of the ocean from near the equator to 25°N. In the Southern Hemisphere, the southeast trade winds cover the eastern two-thirds of the Pacific from 25°S to as far as 10°N in the Northern Hemisphere in the summer. The boundary between these two huge wind systems is called the Intertropical Convergence Zone. The clash of these two systems forces a long line of convection and clouds that is easily seen from space by satellites. Along the equator, these winds, which blow from east to west, pile up warm water in the western Pacific against New Guinea, Indonesia, and the Philippines. The usual situation is that the ocean thermocline (the boundary between the warm, upper layer of water and the cold, deeper layer of water) is relatively deep in the west off New Guinea (about 300 - 500 meters). In the east off Ecuador, it is relatively shallow (about 50 m). As a consequence, there is an enormous pool of 27 - 30°C water in the west, and cold water of 20 - 26°C in the east.   It is the easterly trade winds that kept the status quo. If the trade winds decrease or reverse, the ocean would change through an internal sloshing called an equatorial Kelvin wave. (What is an equatorial internal Kelvin wave? It is a linear wave of elevated or lowered temperature isotherms that run eastward along the equator at a speed of 200 400 km per day. The Kelvin wave has its maximum amplitude at the equator, and extends, in decreasing intensity, up to 1000 km north and south of the equator. When the Kelvin wave reaches Ecuador, it piles up and deepens the thermocline along the coast. The Kelvin wave then propagates toward the poles.) The warm water in the west would

Chapter 3. Global Climate Change

decrease in thickness, but stay the same temperature. The thermocline in the east off Ecuador would deepen from about 50 to 100 m. As a result, two things would happen to the ocean heat budget along the equator. Some of the warm water in the west would be transported to the central Pacific. In the east, the phenomenon of equator upwelling (mixing between the ocean layers) would decrease, and the ocean would no longer be able to diffuse the heat of the sun downward efficiently. The ocean over a large area would warm from about 1 to 3°C.

In 1982, a big El Nino occurred that no one predicted. The atmospheric dust spewed forth by the Mexican volcano El Chicon interfered with the satellite measurements of sea surface temperature. The wind shift that triggered the 1982 El Nino occurred January—April 1982. It sent a huge Kelvin wave across the Pacific, leaving evidence of its passage in its wake. It takes up to three months for the Kelvin wave to cross the tropical Pacific, and one to two months longer for the sun to heat the anomalously deep upper layer. At Christmas Island in May, all the sea birds were content, but by June, millions of adult birds had abandoned their young and eggs. The Kelvin wave had deepened the thermocline, forcing squid and other ocean life to seek cooler, deeper water. Because these food sources were beyond their reach, the birds left to findfood. Most of them never returned. By late fall 1982, a massive El Nino was underway. El Nino typically is regarded as an exaggeration of the seasonal cycle, and therefore should not start to show large temperature anomalies in July or August. However, each El Nino is different in timing and amplitude.

The California El Nino originates in the ocean. The internal Kelvin waves travel up the coast from Ecuador and induce the oceanic disturbances. The Kelvin waves run up the California coast at about 200 km per day. Therefore, a major El Nino Kelvin wave 20,000 km long takes about 100 days to pass a given point on the California coast, such as San Francisco. Afterwards it moves westward very slowly, but will affect the circulation of the California current region for most of a year.

Variability within the natural climate system is historically perhaps the single most fundamental environmental factor affecting the course of human development. Societies, economies, and cultures throughout the world have been developed based in large part on the effectiveness of their ability to adapt to their climate. When temperatures and precipitation patterns depart significantly from historical means, the consequences, especially if unanticipated, can be catastrophic. Examples of the global implications associated with year-to-year variations in the climate include extreme drought such as that experienced in southern Africa in 1991-92; severe flooding, including the recent deluge in the Midwestern United States; and complete elimination of critical sectors of national economies, such as the 1972 -1973 collapse of the Peruvian anchoveta fisheries.

Successful simulation of the mutual evolution of the atmosphere and ocean through coupled modeling has yielded a demonstrated capacity to predict the onset of the ENSO warm events, known to be central to short-term variability in the Earth's climate system. Progress in climate prediction has been stimulated by the development of a variety of models used for simulating ENSO events; by empirical studies that have better defined the global impacts of ENSO; by theoretical studies that have elucidated the underlying oceanic and atmospheric processes accounting for the predictability of ENSO; and by the development of substantially improved observing capabilities in the Pacific for initializing and verifying models under development for ENSO prediction. Compared to the early 1980s when observational techniques were inadequate even to monitor the evolution of an ENSO event once underway, observations are ongoing to detect day-to-day changes in surface winds, sea surface temperature, upper ocean thermal structure, and ocean currents on a basin scale in the tropical Pacific. The capability to forecast the onset of ENSO phenomena, up to a year in advance, is a useful achievement.

Many of the countries most affected by ENSO events are developing countries with economies that are largely dependent upon their agricultural and fishery sectors as a major source of food supply, employment, and foreign exchange. These countries often rely on a regular and predictable climate cycle (e.g., monsoon rains) to provide food and water for their populations. Climate information that can be used by local decision makers to prepare for anomalous precipitation and temperature patterns will provide the means to maintain and enhance their agricultural, fishery, and economic productivity.

Observing Global Change

Many programs relevant to impacts, adaptation and mitigation of global change are underway. Programs on impacts examine effects of global change on the environment and society, including water quality and supply, agricultural systems, and the health and diversity of natural systems. Research in mitigation focuses on the reduction of greenhouse gases and ozone depleting chemicals in the atmosphere by reducing emissions and increasing sinks for these gases. Research in adaptation focuses on developing technologies and modifying practices to cope with climate change and increased UV radiation. Adaptation and mitigation efforts also address costs and feasibility of alternative technologies and their transfer, and provide economic analyses of possible response strategies.

Impacts and adaptation research includes understanding the effects of global change on water supply and quality; review of water utilization and related technologies; understanding the effects of global change on food, fiber and timber supply, and ways to reduce vulnerability in these areas; understanding vulnerability and adaptability of species habitats and maintaining the health and diversity of natural systems; understanding the interactions and effects of global change on humans; and researching human adaptability to change. Continuation of efforts to develop a comprehensive program of land, ocean, airborne, and satellite-based observations on a global scale is critical to understanding and predicting global change. Some recent accomplishments and future plans for the observing program are highlighted below:

In order to advance scientific understanding of the entire Earth system and to develop a deeper comprehension of its components and the interactions among them, it is essential that long-term, comprehensive global observations are collected, archived, and analyzed. A series of polar-orbiting and low-inclination satellites (Space-Based Earth Observing System (EOS)) will provide global observations of the land surface, biosphere, solid Earth, atmosphere, and oceans for a minimum of 15 years. This will greatly enhance the ability to understand and predict the effects of many parts of the Earth system. These include:

> **Geophysical processes**, which have shaped and continue to modify the Earth's surface through volcanism and melting of glaciers and sea ice.
>
> **Hydrologic and dynamic processes**, which control the Earth's temperature and the formation, maintenance, and dissipation of clouds and their interactions with solar radiation.
>
> **Biogeochemical processes**, which contribute to the formation, transport, and fate of trace gases and aerosols and their global distributions.
>
> **Climatological processes**, which govern the interactions of land and ocean surfaces with the atmosphere through the transport of water, heat, mass, and momentum.
>
> **Ecological processes**, which are affected by and/or will affect global change, and their response to such changes through adaptation and adjustment.
>
> **Solar Irradiance**. Because total solar irradiance provides the driving force for the Earth's climate system, it is very important to establish and maintain a long-term record of this energy source. In fact, total solar irradiance variations are suspected of being one of the causes of past global climate changes on decades-to-century time scales.
>
> **Precipitation**. Rainfall is one of the most important parameters which determines climate and, because it is one

of the most variable, climate change. The dynamic and thermodynamic processes which generate rainfall are central to the dynamical, biological, and chemical processes in the atmosphere, in the oceans, and on the land surfaces. Latent heating, the primary internal source of energy for the atmosphere, occurs primarily during the process of condensation of water vapor to liquid water and its subsequent precipitation. It is thus essential to quantify the rainfall (and the energy it releases) in order to understand:

> (1) the dynamics that produce clouds, and circulations and convection that transport and mix minor constituents in the atmosphere; and

> (2) precipitation patterns that cool the land surfaces, nourish the biota, and generate the haline circulations in the oceans. .

Sea Ice Cover. Recent accomplishments have significantly enhanced the ability to monitor global sea-ice cover and to investigate more thoroughly key processes, such as albedo feedback and ocean density modification, that are important to global climate.

Reconciliation with Observations

The world is projected to warm a few degrees (Celsius) over the next century due to the increase in the concentrations of greenhouse gases. Such an increase would be about half of the rise that has taken the Earth from the glacial climate of 20,000 years ago to the warmth of the present. Some studies suggest that the effects on society and the natural environment may be very large and would require significant resources and lifestyle changes to prevent or ameliorate. It is therefore essential to understand the degree of confidence that can be placed in the model results. Models and observations are being tested and compared for a range of time periods. Such comparisons will provide a measure of the accuracy of model representations and an estimate of the confidence of model predictions.

We must diagnose the abilities of global atmospheric models to represent the present climatic state and the observed climatic variations over the recent past, specifically 1979 to 1988. Then the focus will shift to examining the capabilities of ocean-atmosphere models to simulate climate variations and change. To provide more complete data sets for continuing research on model fidelity, four-dimensional data assimilation techniques are being developed and used to combine many kinds of satellite and surface observations to produce carefully checked data products for research, monitoring, and applications. These data will be used to study and test model simulations of the 1986-87 El Nino (warm event) and the La Nina (cold event) of 1988.

Model simulations of paleoclimatic changes are also underway to understand the ability of models to predict very long term climatic change and to understand the sensitivity of climate to past changes in atmospheric composition and radiative forcing [see earlier subsection on Earth system history]. These simulations and related analyses suggest that the modeled estimates of climate change are roughly correct, although problems in understanding the reasons for past climate changes limit the accuracy of the simulations that can be performed.

Of most importance, but hardest to accomplish, is to test model performance against the observed climatic changes of the past two hundred years. This is due in part to there being multiple influences that affect the climate, including: (1) human-induced changes in greenhouse gas concentrations, in ozone concentrations, in the atmospheric loading of sulfate, and biomass aerosols; and (2) natural induced changes in volcanic aerosol loading, in solar irradiance, and, possibly, in ocean circulation. It is especially difficult to construct a model test that can consider all of these factors together for comparison to the changing set of conditions since the start of the Industrial Revolution. The USGCRP agencies are engaged in a continuing effort to formulate an effective approach for carrying out such an analysis.

Predicting Global Change

Chapter 3. Global Climate Change

Models of the Earth system will provide the only rigorous means for developing quantitative projections of the interactions of atmospheric composition, climate, sea level, terrestrial and marine ecosystems, agriculture, water resources, and the effects of human activities. Models and related analysis activities provide the predictive link between the physical Earth system and the human dimensions of global change, including economies, social structure and evolution, and resource use and management. Enhanced activity in global modeling activities build upon recent advances in process studies and observations, with the goals of enhancing forecast capabilities for seasonal to interannual climate variations and strengthening the confidence that can be placed in longer-term climate projections and insights.

Prediction activities contribute to the understanding of the interactions among the components of the Earth system, including: ocean circulation and atmospheric interactions; atmospheric chemistry; potential consequences of emissions of carbon dioxide and other greenhouse gases on climate and of Chlorofluorocarbons and other gases that perturb the stratospheric ozone layer; changes in land use that alter water and biogeochemical balances; changes in biological systems on land and in the oceans, changes in sea level; and changes in how societal systems interact with the climate. Results from model simulations form the basis for the state-of-the-science assessments being conducted both nationally and internationally.

Research on the sensitivity and response of the Earth's climate to natural and human-induced perturbations to the radiation balance, including the effects on the climate due to changes in greenhouse gas concentrations, aerosol loading, and other factors is conducted to support the assessments. Researchers use specific scenarios developed with global models to conduct comparison and evaluation studies of global change projections. Additional modeling studies will include: the climatic effects of aerosols, smoke and {clouds; interannual variations in global and tropical water cycle; land use change and deforestation; and variations in atmospheric chemistry due to human activities, volcanic activity, and natural biogeochemical cycles. Results from model simulations are compared with historical data sets to determine the extent of climate variability that can be explained by the models and their representation of the Earth system.

Accurate predictions of future global change depend on how well models can simulate the many components of the Earth system, including the oceans, atmosphere, land, and biosphere. Global and regional modeling programs are intensifying their efforts to induce vegetation, biological productivity, soil processes, trace gas exchange, hydrology, atmospheric circulation and chemistry, radiation budgets, and ocean circulation. The goal is to provide the basis for these efforts to relate regional scale to global scale patterns and variables (especially precipitation, temperature, and surface roughness) that strongly influence land surface processes and their interactions with the atmosphere.

A high resolution ocean model can accurately simulate mesoscale eddies over the entire globe. The results compare favorably with satellite observations, demonstrating the model's ability to realistically simulate ocean circulation. The description of ocean eddies is a key factor for accurately characterizing fluctuations in the ocean circulation which contribute to climate change on the decadal and longer time scales. A version of this ocean model will be coupled to an atmospheric general circulation model for climate change research in order to contribute to the assessment.

Modeling efforts combine the atmosphere, oceans, land surfaces, and biogeochemistry into an adaptable and flexible system. The biogeochemical component of the models reproduces the seasonal variation in the atmospheric carbon dioxide record. These models are being applied to study the historical evolution of the CO2 concentration. Coupled atmosphere-ocean models are being used to investigate climate variability and climate change on time scales from seasons to centuries. Work sponsored by the USGCRP is progressing on developing realistic representations of land and natural resource interactions, vegetation, and regional land processes for inclusion in Earth system models. Detailed land surface models are being linked to couple ocean-atmosphere system models, and early simulations suggest that significant relationships exist between ocean surface temperatures, atmospheric circulation patterns, and hydrologic conditions over mid-latitude continents. Sea-ice models are also being developed and incorporated into the ocean-atmosphere-land models, to yield more fully coupled climate system models.

Seasonal to Inter-annual Forecasting

Significant success has been demonstrated in forecasting large-scale changes in the sea surface temperature of the tropical Pacific Ocean, especially transitions into the El Nino-Southern Oscillation (ENSO) warming events that dramatically alter precipitation patterns over much of the Pacific basin, including the southwestern United States. This improved understanding will provide both immediate and long-term benefits. On the short-term, the people and economies of the nation and world will benefit from being able to prepare for the consequences of the large natural climate fluctuations caused by ENSO events; on the longer-term, there will be better understanding of whether human activities could affect these fluctuations, which are the largest precipitation anomalies now affecting international agriculture and other human activities.

The development, testing, improvement, and implementation of a quasi-operational experimental forecasting system on seasonal to inter-annual time scales will be extended through the establishment of a multinational network of centers to produce and distribute forecast guidance products. Activities sponsored through this network will include transfer of the products of predictive models to regional application centers around the world so that tailored regional predictions can be provided. Application of regional forecasts on time scales of interest to society are expected to aid in advance planning for agricultural production, resource use, and other societal activities.   This new ability to forecast seasonal to inter-annual anomalies accurately, along with strong indications of the potential for continued progress in predicting anomalies in mid=latitudes, represents a seminal contribution to the understanding of Earth systems processes. Through application of this new technology, this scientific breakthrough can be turned into an effective tool that will contribute to the quality of human life, and to increased economic efficiency.

The global implications associated with seasonal to interannual variations in the climate include potential shifts in the patterns of drought, flooding, and severe storms. These events can have unfortunate social and economic consequences in developed countries, and often have disastrous consequences in developing countries with economies that are largely dependent upon their agricultural sectors as a major source of food, employment, and foreign exchange. Experimental predictions from ENSO models can be used to provide advance warning to decision makers, which allows them to adjust the type and timing of crop planting in anticipation of anomalous climate patterns associated with ENSO events.

Evaluating the consequences

Evaluating the consequences of global change includes determining and interpreting the environmental and societal impacts of global change, and understanding the potential for adaptation and mitigation to ameliorate adverse impacts.

It is proposed to enhance research to improve fundamental understanding of physiological and ecological responses of plants and animals to global changes of climate, atmospheric gas concentrations, and increased UV-B radiation. The relationship must be examined between oceanic physical and biological processes that govern how the recruitment of marine fishes and zooplankton are linked to climatic change. Sensitivity analyses are being conducted to determine the effects of UV radiation on the oceanic carbon cycle in relation to global warming. Methodologies are being developed and validated for estimating the effects of global climate change on lake environmental conditions and fishery resources at a regional scale. The reproductive dynamics of fishery resources, such as sardine, anchovy, and mackerel stocks off the coasts of California, Chile,

Chapter 3. Global Climate Change

Spain, and West Africa are being examined. Research evaluates the potential impacts of climate change at the regional level. Recent modeling studies have successfully simulated winter precipitation at the local scale for a high-elevation, high-water-yielding mountain watershed in western Colorado by coupling regional/local scale atmospheric models to watershed models. These models helped investigate the effects of possible increases in temperature at federally managed reservoirs in western watersheds that might result from a doubling of atmospheric carbon dioxide. The results indicated that the effects of the temperature increases on aquatic ecosystems and fisheries could be significant.

Research on the global and regional effects of climate change on agriculture has shown that, for a moderate scenario of climate change accompanying a doubling of greenhouse gases, the production potential of global agriculture may not be seriously threatened. These results are sensitive to the positive direct effect of carbon dioxide fertilization on plant growth. These results also suggest agricultural production losses may be more severe in developing countries.

Studies determine how atmospheric deposition affects tree growth. The interactions of deposited pollutants with global change parameters (e.g., elevated carbon dioxide) and with biotic stresses (e.g., insect feeding) are also being investigated. These studies have demonstrated that pollutants such as ozone and elevated carbon dioxide influence insect behavior, and hence forest health. Research continually refines ecosystem response models and improve the ability to anticipate ecosystem fluctuations due to climate change.

Contributing research on the adaptation of natural ecosystems to global change includes forest health monitoring, studies on threatened, endangered, and sensitive species, and research into the physiological basis of resistance to drought, ultraviolet radiation, and other stresses for developing crop cultivars that can withstand climate change better than current cultivars. Modeling of the economic impact of climate change scenarios on forest inventories in the southern U.S. is also in progress.

Global change, particularly changes in land use, can have substantial impacts on the preservation of species diversity. Landsat images of Brazil, for example, have shown a large increase in the fragmentation of the habitat, a process that could reduce the number of plant and animal species (biodiversity). The area shown to be severely fragmented, and hence with potentially diminished biodiversity, was more than twice the area actually deforested.

Mitigation Strategies

Mitigation strategy research within the focused USGCRP includes evaluation of CFC substitutes and the environmental implications of the proposed substitutes and their degradation products; the development of models for predicting responses of major agricultural crops to climate change and the development of management tools for ameliorating undesirable effects; and studies on reforestation of presently deforested areas in the tropics.

Research on replacements for Chlorofluorocarbons (CFCs) is focused on understanding potential impacts on humans and the environment. Because the toxicity of many compounds is associated with their metabolism, the metabolism and toxicity of hydrochlorofluorocarbons (HCFCs) and hydrofluorocarbons (HFCs), known collectively as H(C)FCs, are being investigated. The available data indicate that compounds that are rapidly metabolized are more toxic than those that are slowly metabolized. For example, HCFC-132b is very rapidly metabolized and yields metabolites that are very potent inhibitors of the enzymes used by the body to detoxify many drugs and chemicals. As a result, its development has been discontinued. Other research shows that the possibility exists that HCFC-123 may increase susceptibility to hepatitis in sensitive individuals. Finally, computer modeling studies of reactions of H(C)FCs are being conducted. The objective is to develop models that will allow prediction of the rates of metabolism of H(C)FCs and identification of H(C)FCs that are likely to be poorly metabolized and, therefore, have little toxic potential. Results of this research are promising, and the range of compounds to be tested has been expanded.

Chapter 3. Global Climate Change

The biospheric transport and fate of CFC substitutes are also being investigated in order to assess likely future concentrations of these new chemicals in air, water bodies, and soils. Models are being developed to predict the potential ecological impacts of the substituted chemicals and their degradation products.

To ensure a continued abundant supply of food and fiber, research is being conducted on management tools for responding to the potentially undesirable effects of climate change on agricultural productivity. This research involves the development of methods for aggregating plant-scale models to make predictions of impacts at regional scales. Improved predictions of the response of terrestrial ecosystems to changes in temperature, rainfall, solar radiation, especially UV radiation, and changes in carbon dioxide concentrations will enable the development of management strategies for mitigating the adverse impacts caused by these changes.

Programs proposed for greenhouse gas mitigation include the development of more efficient combustion systems which include clean coal technologies, advanced turbine systems for natural gas, and fuel cell systems which operate on natural gas and coal based fuels; alternative energy research, including wind, solar, geothermal and photovoltaics; research on alternative vehicles and fuels, advanced materials research for new transportation technologies; research to develop more energy efficient appliances, space heating and cooling equipment, and energy efficient buildings; research on improving the efficiency of electricity transmission, storage, and distribution reducing the need for new production facilities; and research on reducing greenhouse gas emissions and developing more energy efficient processes for the chemical, petroleum refining, paper, textiles, food processing, and other manufacturing industries.

Problems

1. Research using geodetic observations has demonstrated a correlation between variations in atmospheric and oceanic circulation and variations in Earth dynamics. Show that certain

systematic variations in Earth's angular momentum that have been observed correspond well with El Nino and La Nina events.

2. Polar ice records climate history, and provides the only known continuous, direct record of paleo-atmospheric composition. The analysis of Greenland ice cores have revealed that rapid changes in climate may have occurred over time periods less than a decade, and that these changes are probably associated with switching between stable ocean modes, a switch that may have been triggered by sea-ice conditions.

3. Sand dunes and sand sheets occur extensively on the semi-arid Great Plains. These wind generated deposits are now stable because of the presence of a sparse vegetation cover. Use of a dune mobility index, which incorporates wind strength and aridity factors, shows that increased temperature and reduced precipitation could mobilize the sand deposits throughout a significant area of the U.S. Great Plains

4. The mesosphere is the third major layer of the Earth's atmosphere, directly above the stratosphere and directly below the thermosphere. In the mesosphere, temperature decreases as altitude increases. This characteristic is used to define its limits: it begins at the top of the stratosphere (sometimes called the stratopause), and ends at the mesopause, which is the coldest part of Earth's atmosphere with temperatures below −143 °C. The exact upper and lower boundaries of the mesosphere vary with latitude and with season (higher in winter and at the tropics, lower in summer and at the poles), but the lower boundary is usually located at altitudes from 50 to 65 kilometres above the Earth's surface and the upper boundary (the mesopause) is usually around 85 to 100 kilometres.

Research suggests that the mesosphere and lower thermosphere are exceedingly sensitive to changes in greenhouse gas concentrations in the lower atmosphere—small variations in the greenhouse gas concentrations are predicted to cause large changes in the temperature of these tenuous high-altitude atmospheric regions. Unlike the lower atmosphere which is expected to warm by a few

Chapter 3. Global Climate Change 77

degrees Celsius, a doubling of greenhouse gas concentrations could cool the mesosphere and lower thermosphere by between 10 to 50 °C. Noctilucent clouds are high-altitude clouds that are visible during the short night of the summer. The increase in the occurrence of Arctic noctilucent clouds in the mesosphere over the last twenty years indicates that such a cooling is already taking place. The influence of solar variability on global surface temperature changes may have been significantly underestimated.

6. International population trends and the human condition

> • What are the interactions between population growth/migration and environmental change?

> • What demographic and social factors are particularly relevant for assessing the vulnerability of societies to environmental change?

> • How do the health impacts of environmental change affect consumers and households?

7. Patterns of trade and global economic activity

> • How do world economic growth and international trade patterns affect the use and value of environmental resources such as wetlands, coastal zones, and forest?

> • What effects might large-scale debt-for-nature swaps have on global environmental resources?

> • How will the development and diffusion of technology affect impacts from global changes?

> • What effects might changes such as political and economic liberalization have on the use of environmental goods?

8. Adaptation and mitigation, including environmental resource use and management

- What international mechanisms and processes can be used to build the international coalition needed to stabilize concentrations of greenhouse gases?

- What are the costs and benefits of various policy approaches to influence the use of key resources such as land, energy, water, and coastal zones?

- How do institutional and legal rules affect the use of common property resources?

- Under what conditions has adaptive behavior to environmental stresses been undertaken in the past?

Chapter 4. Air Pollution
"We cannot solve our problems with the same thinking we used when we created them."—Albert Einstein

Pollutants in the Atmosphere
In science and engineering, the parts-per notation is a set of pseudo-units to describe small values of miscellaneous dimensionless quantities. Since these fractions are quantity-per-quantity measures, they are pure numbers with no associated units of measurement. Commonly used are: ppm (parts-per-million), ppb (parts-per-billion), and ppt (parts-per-trillion).

A centimeter (symbol cm) is a unit of length in the metric system, equal to one hundredth of a meter. The prefix "centi" represents a factor of 1/100. A millimeter is a unit of length equal to one thousandth of a meter. Each millimeter is one-tenth of a centimeter. A micrometer or micron is a unit of length equal to one millionth of a meter. A human hair is about 90 microns in diameter. The official symbol for the prefix "micro" is a Greek lowercase mu (μ).

The Earth's atmosphere accounts for only 0.00009 percent of the mass of the planet. Despite its small mass, the atmosphere is vital to all forms of life. An average person breathes about 15,000 liters of air a day but only needs to drink about 2 liters of water. Nitrogen and oxygen are the most important gases in the atmosphere. Air is composed of 78 percent nitrogen. Oxygen accounts for 21 percent, and the inert noble gas argon makes up 0.9 percent of the air. Other trace gases such as carbon dioxide and ozone are important in the way the atmosphere works. Water vapor is another important component and varies from below 0.01 to 5 percent.

The lower part of the atmosphere is called the *troposphere* and extends from the Earth's surface up to approximately 12 km. This is the zone where weather occurs. Human activities introduce pollutants into this zone. Rapid urban and industrial growth has resulted in vast quantities of potentially harmful waste products being released into the air. Wind controls the distribution and dilution of pollution, as well as the direction in which the pollutants move. Rain removes gases and particulates from the atmosphere and transfers them to the land surface or into lakes and the sea. Locally weather perturbations such as inversions can be critical in preventing the dispersal of pollutants and enhancing air pollution in cities such as Los Angeles and Mexico City. Primary pollutants are those actually released into the atmosphere. Carbon monoxide, oxides of nitrogen, hydrocarbons, oxides of sulfur, and particulates account for over 90 percent. Their relative amounts vary. Carbon monoxide, for example, is the most significant pollutant by weight, but when corrected for its tolerance level it becomes less important than some other classes of pollutants. Secondary pollutants are those formed by reactions involving primary pollutants. Photochemical reactions are particularly important. Increased attention by scientists and governments is being given to the health effects of air pollution—identifying specific pollutant sources and their transport in the atmosphere, elucidating exposure-response relationships, and developing air pollution control technology. Despite these efforts, we face a crisis of worldwide air pollution today. This crisis has several aspects:

Natural sources of pollution are volcanoes, radon, biological sources (living, decaying), natural seeps, dust. Man-made sources are carbon monoxide, nitrogen oxides, and hydrocarbons. The primary emissions of sulfur oxides, nitrogen oxides, carbon monoxide, respirable particulates, and metals (e.g., lead and cadmium) are severely polluting cities and towns. Uncontrolled industrial expansion, growing numbers of motor vehicle, and polluting cooking methods and heating fuels contribute to the problem. Poor countries (relying heavily on coal) have higher levels of total suspended particulates than wealthier nations. Even if primary emissions from heavy industry, power plants, and automobiles are reduced, there are new problems coming from pollution by newer industries and from air pollution caused by secondary formation of acids and ozone. The resulting high levels of air pollution cause respiratory disease.

Chapter 4. Air Pollution

Since the atmosphere is dynamic and continually changing, contaminants are transported (sometimes over thousands of miles), diluted, precipitated, and transformed. Pollution, therefore, knows no boundaries or national borders.

Carbon monoxide (CO)

Carbon monoxide is a colorless, odorless, tasteless, flammable gas that forms as a result of poor, or incomplete, combustion. It forms naturally, and the major natural source is the oxidation of methane. About 50 percent of carbon monoxide is from man-made sources and this is generally released in very localized areas. Generally highly populated areas show the highest ambient concentrations with significant variations throughout the day. The single largest source of carbon monoxide is automobiles. At high Ambient levels are generally below 100 ppm. Carbon monoxide reacts with the hemoglobin in the blood and reduces its ability to carry oxygen. Reactions leading to the formation of carbon monoxide include

(1) $2C + O_2 = 2CO$

  $2CO + O_2 = 2CO_2$

(The first reaction above goes 10 times faster than the second one above.)

(2) $CO_2 + C = 2CO$

(3) $CO_2 = CO + O$ (at high temperatures)

Carbon monoxide is emitted mainly from internal combustion engines used in motor vehicles. In the United States, emission controls for new vehicles resulted in a 77% reduction in CO emissions between 1975 and 1981. This decline alone has been responsible for substantial improvement in outdoor CO levels. For example, in the Los Angeles and Orange County basins of California, third-quarter daily maximum hour CO concentrations dropped from approximately 10 ppm in 1970 to 3 ppm in the 1980s. However, because of increased vehicle use and traffic congestion, there remain many localized hot spots in U.S. cities where CO levels still exceed air-quality standards. Exposure can also occur after the burning of coal, paper, oil, gas, or any other carbonaceous material.

The health effects of carbon monoxide have been documented in clinical observations of patients with CO intoxication and in experimental and epidemiological studies of persons exposed to low-level CO. Carbon monoxide is an odorless, colorless gas produced by incomplete combustion of carbonaceous fuels such as wood, gasoline, and natural gas. Because of its marked affinity for hemoglobin, CO impairs oxygen transport, and poisoning often manifests as adverse effects in the cardiovascular and central nervous systems, with the severity of the poisoning directly proportional to the level and duration of the exposure. Thousands of people die annually (at work and at home) from CO poisoning and an even larger number suffer permanent damage to the central nervous system. A sizable portion of the workers in any country have significant CO exposure, as do a larger proportion of persons living in poorly ventilated homes where biofuels are burned.

The pathophysiology of CO poisoning can be conceptualized as anti-oxygen activity, since CO is an antimetabolite of oxygen. Inhaled CO binds strongly to hemoglobin in the pulmonary capillaries, resulting in the complex called carboxyhemoglobin (COHb). The affinity of human hemoglobin for CO is about 240 times its affinity for oxygen. The formation of COHb has two considerable effects: it blocks oxygen carriage by inactivating hemoglobin, and its presence in the blood shifts the dissociation curve of oxyhemoglobin to the left so that the release of remaining oxygen to tissues is impaired. Because of this latter effect, the presence of any level of COHb in the blood, above the background from catabolism of 0.3-0.8%, will interfere with tissue oxygenation much more than an equivalent reduction in hemoglobin from anemia or bleeding. Carbon monoxide also binds with myoglobin to form carboxymyoglobin, which may disturb muscle metabolism (especially in the heart).

The clinical effects of acute CO poisoning vary with the level of COHb, ranging from nonspecific symptoms (headache, dizziness, fatigue) to

Chapter 4. Air Pollution

death. Generally, acute symptoms appear at COHb levels of 10% or greater, and levels above 50% are associated with collapse, convulsions, and death. Because of the profound physiological effects associated with increasing levels of COHb, individuals with certain underlying conditions are at particularly high risk for CO poisoning. These conditions include chronic hypoxemia from lung or heart disease, cerebrovascular disease, peripheral vascular disease, arteriosclerotic heart disease, anemia, and hemoglobinopathies.53 Fetuses and children are more susceptible than adults to CO poisoning.

Chronic, lower-level exposure to CO has also been postulated to accelerate atherosclerotic vascular disease by affecting cholesterol uptake in the arterial wall, though results from animal and *in vitro* studies are conflicting. There is also some evidence that CO may accelerate clot lysis time and may increase platelet activity and coagulation. These alterations in the clotting system could increase the risk for thromboembolism in the heart or the brain.

In addition to sulfur dioxide, particulates, and photochemical pollutants, urban air contains a number of known carcinogens, including polycyclic aromatic hydrocarbons (PAHs), n-nitroso compounds, and, in many regions, arsenic and asbestos. Exposure to these compounds is associated with increased risk of lung cancer in various occupationally exposed groups (e.g., coke oven workers exposed to PAHs and insulators exposed to asbestos). Therefore, populations living near coke ovens or exposed to asbestos insulation at home and in public buildings also may be at increased risk of lung cancer.

In addition to these agents, airborne exposure to the products of waste incineration, such as dioxins and furans, is on the increase in some nations. Though usually present in communities in concentrations much lower than those found in workplaces, airborne dioxins, furans, and other incineration products may still lead to increased lung-cancer risks, particularly in neighborhoods or villages near point sources where their levels may be substantial.

The degree of cancer risk for ambient exposures to these compounds has not been calculated to date.

Concentration of CO in Air	Toxic Symptoms
9 ppm	The maximum allowable concentration for 8-hour period in any year
20 ppm	Typical concentration in flue gases (chimney) of a properly operating furnace or water heater
30 ppm	Earlier onset of exercise-induced angina
50 ppm	Maximum allowable 8-hours work place exposure
75 ppm	Significant decrease in oxygen reserve available to the myocardium
100 ppm	Detectors must sound a full alarm within 90 minutes or less. Slight headache, tiredness, dizziness, nausea after several hours' exposure
200 ppm	Detectors must sound a full alarm within 35 minutes. Headache, tiredness, dizziness, nausea after 2-3 hours, might be life-threatening in long exposures
400 ppm	Detectors must sound a full alarm within 15 minutes. Frontal headaches within 1-2 hours, life-threatening after 3 hours
500 ppm	Often produced in garage when a cold car is started in an open garage and warmed-up for 2 minutes
800 ppm	Dizziness, nausea and convulsions within 45 minutes. Unconsciousness within 2 hours. Death within 2-3 hours.
1600 ppm	Headache, dizziness and nausea within 20

Chapter 4. Air Pollution

	minutes. Death within 1 hour. Smoldering wood fires, malfunctioning furnaces, water heaters, and kitchen ranges typically produce concentrations exceeding 1,600 ppm
3200 ppm	Death within 30 minutes
6400 ppm	Death within 10-15 minutes
12,800 ppm	Death within 1-3 minutes

Nitrogen oxides (NO_x)

From a pollution point of view the most important oxides of nitrogen are nitric oxide (NO) and nitrogen dioxide (NO_2), although nitrous oxide (N_2O) and others can be significant. Natural sources include bacterial action in soils, lightning, and volcanic activity. Man-made nitrogen oxides come mainly from combustion processes including both motor vehicles and stationary sources. Nitrogen oxides play an important role in decreasing visibility (because of the formation of nitrate aerosols), formation of acid rain; and photochemical reactions. The oxides of nitrogen appear to have a generally mild effect on humans with respiratory infections being the most common. The impact on plants comes mainly from their role in generating acid rain.

Hydrocarbons

The major hydrocarbon types are saturated, unsaturated, and aromatic. In the saturated, all four bonds from each carbon atom go to a different atom, i.e. all single bonds. In the unsaturated, there are molecules that contain C-C double bonds. In the aromatic, there are molecules that contain a benzene-type, six-membered ring in their structure. A wide variety of hydrocarbons are released into the atmosphere from both natural and man-made sources. Natural sources are bacterial action, especially important for methane (swamps, marshes, etc.), plants (especially trees), and terpenes.

Man-made are petroleum related emissions, either through refining, transporting, or burning as fuel.

Atmospheric Pollution

The preservation of the environment is paramount. Pollution of our environment is a problem about which nearly everyone gets emotional. The extreme positions on this matter generally pose no solutions; certain trade-offs are necessary. We should be able, with proper planning, to solve these problems. For example, mining must go on more than ever with our present shortages. Yet it must be done with a minimum of environmental disturbance.

In many ways, planet Earth is perilously polluted. There is a deterioration of virtually all aspects of the environment. Air and water pollution may not always be readily apparent, but improper disposal of solid wastes is easily observed. Each American generates 6 pounds of solid trash per day. Farms may add at least 10 times as much waste as all people combined, but fortunately much of this can be recycled. The amount of solid waste generated by industry is undoubtedly huge, but is difficult to estimate. We have covered much of our world with rubbish.   In big metropolitan areas, and to a lesser extent in remote areas, the air carries gaseous and particulate pollutants. Industry is liable, but some studies show that automobiles contribute much more air pollution as does industry.

Our waters are certainly polluted. Industry has so filled some waterways with waste that rivers in industrial centers have on occasion actually burned! Municipalities have not helped the purity of our water. In fact, even a relatively few people in the wilderness can pollute waters. Why are our water and air so polluted? Largely because people have always had the philosophy that the waterways and airways were bottomless sinks for our solid, chemical, and thermal wastes. This bottomless sink philosophy seems especially bizarre, since these water and air sinks have also been the sources of our drinking water and our oxygen. Great strides have been made, but still much time, effort, and money is always needed to keep them clear.

Chapter 4. Air Pollution

As a consequence of rapid urban and industrial growth vast quantities of harmful waste products are being released into the atmosphere. This air pollution affects health, and damages crops, buildings, lakes, and forests. Much air pollution is very localized and is often strongly influenced by local climate and topography. The situation is further complicated because gases commonly react with one another —for example, sulfur dioxide ends up as sulfuric acid and hydrocarbons are oxidized by sunlight to give photochemical smog. Even global atmospheric problems can have a local impact, such as the change in refrigerants required to reduce CFC emissions in efforts to reduce the size of the ozone hole.

In addition to human health effects of air pollution, there are also many other aspects of the problem. For example, damage to ecosystems and agriculture from acid rain, damage to buildings and artworks, and reduced visibility are all attributable to air pollution. Several major types of air pollution are currently recognized to cause adverse respiratory health effects: sulfur oxides and acidic particulate complexes, photochemical oxidants, and a miscellaneous category of pollutants arising from industrial sources.

Photochemical Reactions

Photochemical reactions are reactions in which the initial energy to drive the reaction comes from sunlight. The action of sunlight on hydrocarbons and on the oxides of nitrogen produces secondary pollutants including the photochemical oxidants. Saturated hydrocarbons (e.g. methane) are not involved as much as the unsaturated ones (e.g. terpenes). The photochemical oxidants are compounds that will oxidize materials that are not easily oxidized by oxygen. The most important ones are ozone and peroxyacetyl nitrate (PAN). PAN is a more serious pollutant than ozone. A complex series of reactions operates to generate these compounds in the atmosphere, and when their concentration becomes very high they form photochemical smog.

Photochemical Oxidants

The other two most commonly generated industrial and urban pollutants are ozone and oxides of nitrogen. As opposed to sulfur dioxide, these two substances are produced not so much by heavy industry as by the action of sunlight on the waste products of the internal combustion engine. The most important of these products are unburned hydrocarbons and nitrogen dioxide (NO_2), a product of the fixation of atmospheric nitrogen with oxygen during high-temperature combustion.

Ultraviolet irradiation by sunlight of this mixture in the lower atmosphere results in a series of chemical reactions which produce ozone, nitrates, alcohols, ethers, acids, and other compounds that appear in both gaseous and particulate aerosols. This mixture of pollutants constitutes the smog that has become associated with cities with ample sunlight (such as Los Angeles, Mexico City, Taipei, Bangkok) and with most other urban areas in moderate climates. Ozone and nitrogen dioxide have been studied extensively in both animals and humans. Both gases are relatively water insoluble and therefore reach lower into the respiratory tract. Therefore, these gases can cause damage at any site from the upper airways to the alveoli.

Ozone
The most important photochemical oxidant, and the one most commonly reported, is ozone (O_3). Natural background values are between 0.005 and 0.05 ppm (10 - 100 mg/m^3) but in polluted areas can rise into the 0.15 to 0.40 range. In a typical year, a value of 0.10 ppm is exceeded in Los Angeles on about 100 days and in Pasadena on about 170 days. The photochemical generation of ozone from primary pollutants takes time, and the maximum concentration of ozone may be as much as 10 to 60 km downwind from the original source of pollution.   The long term health hazards of ozone are not known, but short term exposure causes eye irritation, as well as headaches, coughs, sore throats, and chest discomfort at 0.10 to 0.15 ppm levels. Ozone produces cracks in rubber, and degrades textiles and exterior painted surfaces. It can kill trees, and tomatoes and tobacco are especially sensitive to it. Since ozone promotes the

Chapter 4. Air Pollution

oxidation of sulfur dioxide to sulfur trioxide it increases smog and regional haze.

Exposure to Ozone and Nitrogen Dioxide
Clinical research on the acute effects of exposure to ozone reveals symptoms and changes in respiratory mechanics at rest or during light exercise after exposure to 0.35 ppm. In exercising subjects, similar effects will occur with exposure to 150 ppb, a level currently found in urban air. There is also recent evidence that the upper-airway (nose) inflammation induced by ozone correlates with the lower-airway response, as is demonstrated by simultaneous nasal and bronchoalveolar lavage. Animal studies have clearly documented severe damage to the lower respiratory tract after high exposure, even briefly, to ozone. The pulmonary edema noted is similar to that noted in humans accidentally exposed to lethal levels. Animals exposed to sub-lethal doses develop damage to the trachea, the bronchi, and the acinar region of the lung.

Levels of ozone that exceed the federal standard (120 ppb) are frequent in cities in Southern California and in the Northeast. Large air masses with ozone concentrations that exceed the federal standard have been observed under stable meteorological conditions that can last several days in the Northeast. High concentrations of ozone are most often observed in the summertime, when sunlight is most intense and temperatures are highest—conditions that increase the rate of photochemical formation. Strong diurnal variations also occur, with ozone levels generally lowest in the morning hours, accumulating through midday, and decreasing rapidly after sunset. While ozone concentrations may be elevated in outdoor air, they are substantially lower indoors. The lower indoor concentrations are attributed to the reaction of ozone with various surfaces, such as walls and furniture. Mechanical ventilation systems also remove ozone during air conditioning. For these reasons, staying indoors or closing car windows and using air conditioning are generally recommended as protecting against exposure to ambient ozone.

Studies of the effect of nitrogen dioxide on healthy human subjects have demonstrated results similar to those found with ozone, with the exception that the concentration of nitrogen dioxide shown to produce mechanical dysfunction is above the concentrations of ozone noted in pollution episodes (i.e., 2.5 ppm). In animal studies, similar damage to alveolar cells is noted after exposure to relatively high concentrations of nitrogen dioxide. In any case, epidemiological data show that exposure to photochemical oxidants, particularly ozone, can cause bronchoconstnction in both normal and asthmatic people.

Oxides of sulfur
The most important sulfur gases in the atmosphere are sulfur dioxide (SO_2) and sulfur trioxide (SO_3), with the former dominating. Although sulfur dioxide is formed naturally by the oxidation of hydrogen sulfide the main manmade sources are combustion of coal and oil and industrial processes such as smelting. In the atmosphere sulfur dioxide is oxidized to sulfur trioxide which then reacts with water to form sulfuric acid. Aerosols of acid droplets commonly develop. In humans the sulfur oxides cause eye and throat irritation. They also damage or kill leaves and lead to soiling and corrosion of buildings.

Sulfur Dioxide and Acidic Aerosols
Sulfur dioxide (SO_2) is produced by the combustion of sulfur contained in fossil fuels such as coal and crude oil. Therefore, the major sources of environmental pollution with sulfur dioxide are electric power generating plants, oil refineries, and smelters. Some fuels, such as soft coal, are particularly sulfur-rich. This has profound implications for nations such as China, which possesses 12% of the world's bituminous coal reserves and depends mainly on coal for electric power generation, steam, heating, and (in many regions) household cooking fuel.

Sulfur dioxide is a clear, highly water-soluble gas, so it is effectively absorbed by the mucous membranes of the upper airways, with a much smaller proportion reaching the distal regions of the lung. The sulfur dioxide released into the atmosphere does not remain

Chapter 4. Air Pollution

gaseous. It undergoes chemical reaction with water, metals, and other pollutants to form aerosols. Statutory regulations promulgated in the early 1970s by the U.S. Environmental Protection Agency under the Clean Air Act resulted in significant reductions in levels of SO_2 and particulates. However, local reductions in pollution were often achieved by the use of tall

A concentrations of SO_2 as low as 0.4 ppm can cause symptomatic bronchoconstriction when inhaled by exercising asthmatics. These levels may be encountered in polluted urban air, and exposure to these levels for even several minutes is enough to induce bronchospasm. So, sulfur-dioxideinduced bronchoconstriction is not necessarily prevented by the current EPA standard of 0.14 ppm as a 24-hour maximal average, as this standard does not set limits for maximal concentrations over shorter periods. In addition to the acute bronchoconstnctive effects of sulfur dioxide, there is epidemiologic evidence for chronic airway obstruction in persons exposed to elevated levels of SO_2.

Particulates
Particulates are frequently emitted by the same sources as sulfur dioxide. They are a wide range of finely divided solids or liquids dispersed in the air. Particulates come from natural sources such as forest and grassland fires caused by lightning strikes, combustion processes, and industrial activities. Particle sizes range from 0.1 to about 25 mu in diameter (1 mu equals 1 micron). Total suspended particulates (TSP) are determined by weight. Particulates, along with sulfur dioxide, are regarded as the traditional pollutants that cause smog (i. e., smoke/fog).

Particulate air pollution is closely related to SO_2 and aerosols. The term usually refers to particles suspended in the air after various forms of combustion or other industrial activity. Air pollution is characterized by high levels of particulates, sulfur dioxide, and moisture. Studies have shown that particulate air pollution *per se* is associated with increases in daily mortality both in heavy smog episodes and in lower pollution levels. Total suspended particulate counts are associated with increased daily mortality. High particulate levels (both indoor and outdoor) result in adverse health consequences such as high death rates in children with acute respiratory disease. After infant diarrhea, acute respiratory disease is believed to be the major cause of death in children under the age of 5 in the developing world. Air pollution, both indoor and outdoor, appears to play a central role in this epidemic.

Lead
Eighty percent of the lead in the atmosphere comes from gasoline with the other 20 percent contributed by smelting, painting, and other industrial processes. Lead has been added to gasoline as an anti-knock compound for over 50 years. Near highways the lead content of plants is high but it decreases with distance away from the road.   It is used mainly in the form of alkyllead (tetraethyllead and tetramethyllead). The increasing use of unleaded gasoline has decreased the amount of lead entering the atmosphere, but unleaded gasolines are more expensive and use of leaded gasoline in a car not designed for it disables the catalytic converter. As a consequence carbon monoxide and hydrocarbon emissions rise by a factor of 4-8.

Acid Rain
Rainwater normally has a pH of about 5.6. It is slightly acid because carbon dioxide dissolves and makes the weak carbonic acid. Human activities appear to be causing a dramatic increase in acidity on a local and regional level by introducing oxides of sulfur and nitrogen into the atmosphere. These are converted into sulfuric and nitric acids that return to earth with rain and other forms of precipitation. Acid rain has been linked with the depletion of fish populations in lakes as pH decreases. In terrestrial ecosystems increasing soil acidity can decrease nutrient availability, mobilize toxic metals, leach important soil chemicals, and change species composition.

Indoor Air Pollution
No description of the health effects of air pollution is complete without a discussion of indoor air pollution. There has been a failure to address indoor air pollutants. The descriptions above generally have referred to community (ambient) air pollution. Moreover, as

mentioned above, much of what we now know about the health effects of airborne toxins derives from the study of workers in factories and mines—both of which are indoor environments. However, today, both the common and the scientific use of the term indoor air pollution refer to homes and non-factory public buildings (modern office buildings, hospitals, etc.). Most indoor environments, whether they be traditional village homes or tightly sealed high-rise office buildings, have air pollutant sources. Pollution can come from heating and cooking combustion, pesticides, tobacco emissions, abrasion of surfaces, evaporation of vapors and gases, radon, and microbiologic material from people and animals. High concentrations of pollution in indoor settings in either the developed world or the developing world can be associated with mucous-membrane irritation, discomfort, illness, and even death.

Although indoor air pollution has increased substantially in the industrialized nations because of tighter building construction and because of the widespread use of building materials that may give off gaseous chemicals, indoor air pollution is a particular problem in the poor communities of developing countries. Wood, crop residues, animal dung, and other forms of biomass are used by approximately half the world's population (2.5 billion people) as cooking and/or heating fuels, often in poorly ventilated conditions. This leads to high exposures to such air pollutants as carbon monoxide and polycyclic aromatic hydrocarbons, particularly for women and children.

There are strong indications that indoor air pollution is associated with acute respiratory infection in infants and in children under 5 years. In view of the large numbers of children from developing countries who die from respiratory infection each year, this is a high priority for public health research and action. Approaches to controlling air pollution have varied from nation to nation. For example, in Great Britain the approach was to recognize that outdoor urban air—not specified as to pollutants—was unhealthy.

The approach in the United States has been different. Congress passed a Clean Air Act in 1970 which required the Environmental Protection Agency to define air quality standards allowing a margin of safety to protect the public's health. This required identifying specific pollutants and setting standards for each. The pollutants chosen as criteria were sulfur dioxide, carbon monoxide, nitrogen dioxide, ozone, lead, and total suspended particulates. The assumption underlying this approach is that scientific research would allow the identification of a threshold concentration for each pollutant, below which adverse health effects would not occur. Even with the inherent problems of this assumption, standards were set for all six pollutants. Therefore, the federal government prescribed procedures and standards for the states to follow. The laws dictate that polluters file permits and maintain documentation, and that failure to comply will result in potential fines, loss of operating permits, liability for cleanup, and civil or even criminal penalties for cleanup and/or injury. There are obvious instances where compliance with the law still does not ensure health protection against air pollutants, or ensure environmental protection. For example, in the construction of new highways, the environmental impact review process assesses compliance with air quality standards at outdoor locations. Ignored in the assessment for the highways are other important impacts, such as the associated industrial and construction projects, the increase in miles traveled in vehicles (and the consequent emissions), and the increased general energy expenditure for utilities such as electricity.

The limitations of the current U.S. approach to air pollution can be briefly summarized as follows:

> 1. Ozone and lead are clearly not adequately controlled under the current standards (section 109 of the Clean Air Act). Also, there is inadequate control of acidity inherent in the SO_2 limits. Moreover, in section 112 (which addresses carcinogens), the U.S. government's approach has been not to force new technological development of controls, but rather to use the least expensive available methods.

> 2. There has been a failure to adequately address non-criteria airborne pollutants that affect human health through indirect mechanisms— i.e., CO_2 and chlorofluorocarbon emissions as

causes of global warming and stratospheric ozone depletion, respectively.

The problem of controlling air pollution is indeed a pressing one. Atmospheric pollution has now reached a level that threatens not only the health of entire populations but also their survival. Various national regulatory approaches have not, so far, been up to the task of controlling pollution on a global scale. Air pollution is a growing, global problem. Only global approaches will succeed in controlling it. There are probably more chemicals in a typical house today than there were in a well-equipped chemistry laboratory a century ago. Many of the synthetic products in everyday use enhance the comfort and convenience of modern living. They are used in furnishings, appliances, and even cosmetics. They are also used for cleaning, as solvents, and for killing pests (both in the house and in the garden).

Formaldehyde
Formaldehyde is a colorless gas first synthesized in 1859 and used to make the first synthetic plastic in 1909. Applications developed rapidly. Much of it is used by the forest products industry as a bonding agent particle board, hardwood plywood, medium-density fiberboard, and softwood plywood. In addition newsprint, waxed papers and grocery bags also contain formaldehyde, as do many common household products including carpets, fabrics (treated to resist mold, fire, and crushing), paints, waxes, polishes, glues, adhesives, molded plastics, insecticides, fumigants, cosmetics, and household cleaning products.

During the 1970s formaldehyde was used in the form of ureaformaldehyde foam (UFFI) and pumped into wall cavities of houses to provide thermal insulation. Nearly half a million houses and mobile homes were insulated in this way, but unfortunately it caused a large number of health problems, including ear, nose, and throat irritations and skin rashes, and in more severe cases, dizziness, vomiting, memory loss, and nosebleeds. High concentrations of formaldehyde forced some people to abandon their homes and in 1982 the Consumer Product Safety Commission banned UFFI. The ruling was appealed by the Formaldehyde

Institute and overturned. While there appears to be no problem if UFFI is correctly mixed and installed it is rarely used today.

Problems arise with formaldehyde because it is slowly and continuously released from materials by a process called outgassing. For example, in particle board that is glued together with urea-formaldehyde resin the residual, unreacted formaldehyde is slowly released. This release steadily diminishes with time but can be quite intense for the first six months, and still remain significant for years. Low-level outgassing continues indefinitely. Formaldehyde is classified as a probable human carcinogen and is reported to cause cancer in rats. Human risk is not known but reactions such as memory loss are common.

Volatile organic compounds

Formaldehyde is one example of a volatile organic compound (VOC) that easily evaporates into homes or offices. Others include easily vaporized organic compounds that are released by outgassing. Many of the materials used in building interiors and cars are plastics. These represent a wide range of polymers build up from repeated small units called monomers together with plasticizers. Slow release of these compounds can contribute to the interior VOCs. The new car smell is caused by outgassing plastics.

Asbestos

Asbestos has been known and used since ancient times, and the name itself comes from the Greek word for inextinguishable. The term asbestos is applied to a variety of different fibrous minerals all of which are obtained by mining natural materials (5 million tons in the mid 1980s). Asbestos fibers are fine, strong, flexible, and can stand temperatures up to 400 °C. They are also waterproof, resistant to friction, and sound adsorbent. This combination of properties has lead to the use of asbestos in thousands of products, including insulation and fireproofing. It is most likely to be found in buildings constructed before the 1950s .   In the U.S. 96% of the asbestos used is chrysotile (white asbestos), 2% is amosite (brown asbestos), and 2% is crocidolite (blue asbestos).

The health hazard posed by asbestos was first recognized in a French textile mill in the early 1 900s where many of the workers were developing lung disease. Other cases were reported in miners and asbestos workers, particularly those that were smokers. Inhaled asbestos fibers accumulate in the lungs and can cause asbesteosis (a scarring of the lungs) or mesothelioma (a rare cancer of the lining of the lungs). It is only the crocidolite that has been associated with mesothelioma. OSHA limits workplace exposure to 0.2 fibers/cm3/8hour period. There are no limits for homes. Detailed studies show that it is only crocidolite that is a major health hazard and this accounts for only 2% of the asbestos used in the U.S. In the San Francisco area natural amphibolite weathering puts asbestos into the air with a concentration 1000 times greater than the EPA limit for schools. No above average number of cases of cancer have been reported in this area.

Radon
Radon is a colorless, odorless, inert gas that is produced by the radioactive decay of uranium-238. Radon-222 itself is radioactive with a half life of 3.82 days, and decays through a series of alpha and beta emissions and eventually gives the stable isotope lead-206. Along the way it generates polonium and bismuth, both of which are carcinogenic.   Natural radon concentrations will be highest in areas of high natural uranium content, and this includes granites, phosphatic shales, some organicrich shales, and uranium ores. It is continuously seeping to the Earth's surface and may become trapped in houses, especially if the basements are inadequately sealed. In some areas it is dissolved in the ground water.

Radon itself is not a major radiation threat, but if the radioactive bismuth, lead, and polonium daughter products form from radon gas in the lungs they can lead to lung cancer. Lung cancer is the only health hazard that has been identified, and estimates range from 5000 to 30,000 deaths/year from this source. The presence of radon can be tested either by activated charcoal adsorption (1 week) or by alpha-track devices (approx. 4 weeks) which give more accurate long-term results. Remediation can be effected either by preventing access or by increasing ventilation.

Lead

Lead was widely used in building many older houses. Lead water pipes were commonly used in the past and even copper pipes were joined with solder containing lead. Until fairly recently many paints also contained lead. As painted surfaces wear, for example on the edges of doors, lead-bearing dust is formed. One must be careful when old lead-based paint is being removed by sanding or scraping. The adverse health effects of lead have been well documented. At high levels it affects the central nervous system, and may cause mental disorders, urinary tract disorders, anemia, bone damage, and reproductive dysfunctions. Young children are ewspecially at risk.

Weak electromagnetic field (EMF)

EMF denotes **weak, extremely low frequency, electromagnetic field**. Such fields are an inevitable consequence of an electric current flowing through a wire. They may be generated by many household appliances and by power transmission lines. Although it is officially claimed that there is no conclusive evidence linking EMF to health problems, there is a growing body of evidence that has linked them to leukemia, various cancers, and depression, among others. These fields are measured in units of milligauss, with 0.5 milligauss being a typical background value. Fields close to household appliances can be much higher, and in a kitchen with all appliances operating can exceed 50 milligauss. Activists claim 2 milligauss as a safe limit. EMFs can be measured by a device called a gaussmeter. Fields decrease with distance from the source in accordance with the inverse square law. This shows that if the distance doubles then the field drops by a factor of 4 (i.e. 2 squared). As a general guide it is a good idea to stand 3 feet or more from microwave ovens and TVs.

Thermal Pollution and Noise

Burning of fuel generates heat. Waste heat can be evolved directly into the atmosphere, or released indirectly through cooling ponds and rivers. In cities many heat producers are clustered together and produce a so-called heat island effect. This **thermal pollution** may change the local climate. Changes in the nature of surfaces in cities, such as paving parking lots, can also affect the heat balanc.

Chapter 4. Air Pollution

Noise pollution refers to unwanted sound. Sound (and noise) moves through the air as a series of pressure waves made up of successive compressions and rarefactions. Deforming the air as the wave passes takes energy, and we can describe loudness in terms of the energy involved — for example, a symphony orchestra is as loud as a 10-watt siren. However, the range of energies involved is very large and it is more convenient to describe sound intensity on a scale that increases by a factor of 10 for each unit. The unit used is the decibel (dB). In going from 40 dB (soft music) to 50 dB (a low conversation) the energy in the sound increases ten-fold.

Chapter 5. Soils and Solid Waste

"We know more about the movement of celestial bodies than about the soil underfoot."—Leonardo da Vinci

Food production

Food production on land could be increased by increasing the area under cultivation. For various reasons, more than 80 percent of earth's land cannot be cultivated; 9 percent is in crops, and another 9 percent has potential. In Europe and Asia, 80 to 90 percent of arable land is cultivated already, and in much of South America, Africa, and Australia either the soil is of low quality or the climate is too dry. And there are problems with opening up new land to cultivation. For example, clearing tropical forests and farming the land results in its transformation in just a few years to hard brick-like soil. Clearing non-tropical forests would result in more acreage for crops, but forest soils are soon depleted of their nutrients, and other ecological problems commonly result. To utilize semiarid or arid areas would require tremendous quantities of water, which would generally have to come from expensive desalting of ocean water. The best soils are already in use. Yet if all modern technology and knowledge could be applied, with cost not a factor, earth's arable land area could be doubled and the crops might feed a population several times that of today. But why push earth and mankind to such limits?

An allied problem is that while we may strive to put more land under cultivation, some of the best lands are being taken out of production by urban sprawl. Over one-third of California's best land is under lawn and concrete. Crops grown on soil are directly related to climate.   Climate is the most important factor governing the processes of chemical weathering of rock and hence the formation of soil. Regions with moderate rainfall and moderate temperatures—the temperate zone—have the best soils. In regions of low rainfall or low temperatures, little chemical weathering occurs. In regions of high rainfall and high temperatures—the tropics—there is too much and too rapid chemical weathering and the nutrients are removed from the soil. In some areas, such as Java, more than a foot of rich soil

Chapter 5. Soils and Solid Waste

can form on glassy volcanic ash in only 50 years, but in most regions, the soil forming process takes thousands of years. Plants, and thus the creatures that eat them, get their nutrients from certain clay minerals and from decaying vegetation; thus, some climatic regions produce nutritious plants and other regions produce less nutritious plants. For example, corn, which grows in the temperate zones, is a low-bulk and high-nutrition food, whereas the banana, which grows on poorer subtropical to tropical soils, is a high-bulk and low-nutrition food. The three most important crops in the world are rice, wheat and corn, which are all relatively high in protein. Wheat and rice provide 40 percent of all human food energy. Potatoes, plus many crops grown in the tropics, are low-protein foods. Many food experts are now saying that people, rather than animals, should eat the grains. Cattle, sheep, and hogs use 7 calories of food for each calorie of meat they produce. Instead of raising animals to eat, the human consumption of the grains would be much more efficient in terms of calories. Of course, animals could be grazed on land that cannot be used for growing crops.

However, all soils are vulnerable; humans can really spoil the soil by overuse without replacing the nutrients. Even with fertilization, some studies indicate that the fertility of Iowa's soil is slowly decreasing. Adding too much fertilizer can result in the carrying off of many of the nutrients by running water, causing lakes to be choked with plant growth. Thus, in attempting to solve one problem, we generate others. Can the green revolution of new high-yield plants help? Yes, the results have been encouraging. In crop year 2016/2017, a total of approximately 2.62 billion metric tons of grain were produced worldwide. The most important grain was corn, based on a production amount of about 1.05 billion metric tons in that year. However, there could be related dangers. Stocks of seeds of the long-used, older varieties are stored. The reason is that these stocks are kept for insurance. If the new, high-yield varieties prove to be susceptible to insects and diseases which the older varieties resisted, we could go back to the older varieties. Also ecological factors are involved: the green revolution required a 300 percent increase in pesticides, as well as the use of much more fertilizer and energy.

All told, it takes 68 calories of fossil fuel to produce one calorie of pork, and 35 calories of fuel to make one calorie of beef. Roughly twenty-five times more energy is required to produce one calorie of beef than to produce one calorie of corn for human consumption. The maintenance of a high meat diet could become expensive. New technology and equipment may not be the answer either. So, can we afford to fertilize and irrigate poorer lands and to further develop our technology? If so, maybe we can significantly increase our acreage, but the food will be costly and the limitations are real. Will such efforts do irreversible damage to the croplands, water resources, forests fisheries and other ecological resources on which life ultimately depends? Nearly 80% of the world's deforestation is due to increased demands. The regions of the world are interdependent. Various regions must draw on other regions for resources.

Historically human inventiveness has succeeded in raising crop yields and coping with diminishing resources. Technology has greatly increased food production. For the last 100 years this approach has seemingly been correct. Products of human ingenuity, like the science-based Green Revolution, led to the development of high yield grains. However, no one can say what the carrying capacity of the earth may be, since it depends on unknown changes in technology and on the unknown ability of economies to substitute new resources for ones that are running out.

Environmental change may be abrupt and nonlinear. Continuous increases in fishing cannot be maintained without having a serious impact on the fish population. A threshold or crucial point is reached at which the reproduction of the fish is interfered with and the fish population collapses. Between 1900 and 1950 sardines were in huge abundance in California's Monterey Bay and supported a large industry. Fished out, they disappeared suddenly, and they have not returned. Population scientists also make an important distinction between renewable and nonrenewable resources. For the near term

Chapter 5. Soils and Solid Waste

there seem to be sufficient nonrenewable resources. The prices paid for most metals, for natural gas, and for oil have declined or remained low in the last half-century. Ironically, it is the renewable resources—lumber, water, and fish, for example that in our increasing numbers we are exhausting. A disturbing report published by the World Wildlife Fund in 2015 found that the world marine vertebrate population declined by 49 percent between 1970 and 2012. Populations of some commercial fish stocks, such as a group including tuna, mackerel and bonito, had fallen by almost 75 percent, according to the study. It also found that local and commercial fish populations have been cut in half, tropical reefs have lost nearly half of their reef-building coral, and there are 250,000 metric tons of plastic in our oceans.

Agricultural intensification does increase food production, but there are limits to this solution. Productivity gains from intensive agriculture are limited by availability of energy and fertilizer, water supplies, and the loss of topsoil. There is a phenomenon of land degradation that scientists term **desertfication**. Desertification is the impoverishment of land by human activities. Its causes are overgrazing, over cultivation, salinization, and deforestation. Desertification occurs when the carrying capacity—the number of people a given area of land can support—is exceeded. It turns out that firewood resources may be the limiting factor for carrying capacity in some temperate and hot climates. Natural cycles of drought further lower a region's carrying capacity and hasten desertification. Overgrazing of marginal lands by cattle is a major cause of desertification. Cattle husbandry contributes to rain forest destruction, global warming, water pollution, and water scarcity.

Soils

The Earth is made up of layers. Geophysical methods have shown that at the center is the core (which is mostly iron and nickel); next comes a silicate mantle (divided into an Upper Mantle and a Lower Mantle); and, finally, a very thin outer skin, only 10 to 60 km thick, called the crust. The uppermost part of the crust is the only material we can observe directly in place.

Rocks are made up of one or more minerals, each of which has its own particular chemical composition. 98% of all minerals are made up of only 8 elements. Except for a few common minerals, identification in hand specimens is difficult and usually microscopic examination is needed.

ABUNDANCE OF MINERALS IN THE EARTH'S CRUST

MINERAL	VOLUME PERCENT
PLAGIOCLASE	42
POTASSIUM FELDSPAR	22
QUARTZ	18
AMPHIBOLE	5
PYROXENE	4
BIOTITE	4
MAGNETITE, ILMENITE	2
OLIVINE	1.5
APATITE	0.5

CARBONATES
Calcite (calcium carbonate), important in limestones.
Dolomite (magnesium calcium carbonate).

SILICATES
Feldspars are the commonest minerals.
Quartz, micas, pyroxenes, amphiboles and olivine are important rockforming silicates based on the SiO4 unit.

OTHER MINERALS
Limonite is a hydrated iron oxide that occurs widely in small amounts and commonly gives a reddish-brown color to rocks (iron staining). Gypsum or anhydrite is a calcium sulfate that is the major source of sulfate in subsurface waters.
The 3 commonest sulfide minerals are galena (lead sulfide), sphalerite (zinc sulfide), and pyrite (iron sulfide).

Weathering refers to the break-down of rocks when they are exposed to an environment that is different from the one in which they formed

Chapter 5. Soils and Solid Waste

Mechanical weathering represents the physical breakdown of the rock e.g. by ice or wind. Chemical weathering represents the change in composition of the rock due to chemical reactions oxidation hydrolysis carbonation.

Soils interact with both water and the atmosphere and play a key role in overall environmental processes. Soils can receive pollutants that come as airborne particulates or from percolating waters. Agricultural practices that apply large quantities of fertilizers, pesticides, and herbicides, coupled with irrigation, introduce a wide range of materials into soils and these may be retained, modified, and eventually remobilized. Locally soils may be contaminated by industrial activities including mining, petroleum production, waste disposal, and radioactive waste storage.

Agriculture

Agriculture gives rise to several sources of contamination. Fertilizers produce nitrogen, phosphorus, potassium, all a potential for eutrophication. Pesticides and herbicides (chlorinated, organophosphates). There is soil erosion and sediment transfer. There is slow biological degradation and things remain in environment for a long time. A pilot project is underway in a tropical area severely affected by deforestation and consequent erosion to study the restoration of the fertility of degraded soils and the improvement of forest cover. The system combines small-scale tree farming, horticulture, fodder production, and animal husbandry. Future plans include the expansion of the area to explore economies of scale, and to test new tree and crop species.

Municipal Solid Waste

It is important to recognize when a site is contaminated and when cleanup is required. In some cases it may be obvious, for example if there are major spills, mine spoil tips, leaking surface facilities, etc. In other cases contamination may be much more subtle and it may need sophisticated analytical techniques to document the presence of a few ppm of toxic substances.

The solid waste problem has three facets: source, collection, and disposal. The amount of solid waste can be lessened by using new concepts in packaging, use of natural resources, and evaluation of planned obsolescence. However, economic factors often discourage this. We may need to create a new economy with regard to productivity, obsolescence, and waste before we can begin to be at peace with our ecosystem and have any hopes of long-term survival. There are relationship between solid wastes and human disease. Improper solid waste disposal is a true health hazard because of the human diseases that are associated with solid wastes. Two transmitters of human disease in regard to solid wastes are rats and flies. The overriding criterion when municipalities decide on a method of disposal is cost, not health! What is the cheapest way to get rid of this stuff?

Solid wastes are classified into municipal solid waste (MSW), hazardous waste and radioactive waste. Municipal solid waste (MSW) consists of solid and semisolid materials discarded by a community. The fraction of MSW produced by a household is called refuse. Refuse includes food waste, plastics, packaging, and discarded household items. The components of refuse are garbage(food wastes and other organics); rubbish (glass, tin cans, and paper); and ashes, still a problem where coal is used for heating homes. Lastly, trash refers to such larger items as tree limbs, old appliances, and so forth, which are not normally deposited into garbage cans. Many of the products manufactured for household use eventually become municipal refuse.

Ground water and drinking water can be contaminated by leachate from solid waste. Leachate is formed when rain water collects in landfills and stays in contact with the material long enough to leach out and dissolve some of its chemical and biochemical constituents. Leachate can be a major ground water and surface water contaminant, particularly where there is heavy rainfall and rapid percolation through the soil.

Solid waste is collected by trucks, usually equipped with hydraulic rams to compact the refuse to reduce its volume. On the average, about 80 percent of the total cost of solid waste management is spent on collection. In many instances, the operation is one of inefficiency. New devices and methods are continually being introduced to cut collection cost.

Garbage grinders reduce the amount of garbage in refuse. If all homes had garbage grinders, the frequency of collection could be cut in half, since the twice-a-week collection in most communities is necessary only because of the rapid decomposition of the garbage component. Obviously, garbage grinders put an extra load on the wastewater treatment plant, but sewage is relatively dilute and ground garbage can easily be accommodated both in the sewers and in treatment plants.

Pneumatic pipes have been installed in some small communities, mostly in Sweden and Japan. The refuse is ground at the residence and sucked through underground lines. Pneumatic pipes may be the collection method of the future.

Transfer stations are found in almost all larger communities. A typical system involves several stations scattered around a city to which ordinary collection trucks bring the refuse. The objective is to have workers spend more time collecting and less time traveling. At the transfer station, bulldozers cram the refuse into large cans, which in turn take the material to the ultimate disposal site. *Route optimization* can result in significant savings to a city. Several computer programs are available for selecting the least-cost routes and collection frequencies. Such optimization techniques have resulted in increased effectiveness and lower cost of refuse collection.

Disposal Options

Until the last few decades, municipal solid waste was not a problem. As cities grew land for municipal dumps became scarce and expensive. The amount of solid waste per household has vastly increased. Home incineration of trash and backyard burning used to be allowed. However, concern about air and water pollution eventually resulted in widespread prohibition of backyard burning, even of leaves and grass clippings and de-emphasis of municipal incineration.

First, the solid waste must be collected. Then it must be either disposed of or recycled. All of these—collection, disposal, and/or recovery—form a part of the total solid waste management system. The disposal of solid wastes involves the placement of the waste so it no longer affects people. One method is to assimilate the residue so it can no longer be identified in the environment (for example, fly ash from an incinerator) or to hide the wastes well enough so they cannot be readily found. Another method is to process the solid waste so that some of its components can be recovered, a procedure popularly known as recycling. Before disposal or recycling, however, the waste must be collected.

Litter is a ubiquitous environmental problem. Litter is unsightly, unhealthy (as a breeding ground for rats and other rodents) and damaging to wildlife. Deer and fish, attracted to aluminum can pop tops, ingest them and die in agony. Plastic sandwich bags are mistaken for jellyfish by tortoises and death results. The availability of trash cans seems to be an important factor influencing littering. A person will walk or ride only a short distance out of the way to deposit waste into a litter can. Increasing the availability and the frequency of maintenance of litter deposits can have a marked effect on litter. Another assault on the litter problem is restrictive legislation. Some states prohibit the use of pop-top cans and discourage nonreturnable glass beverage bottles.

Solid Waste Disposal

Refuse can be disposed of either as is or after suitable processing. This processing may be thermal or physical, and is performed only for the purpose of converting refuse to a more readily disposable form and not as a method of energy or materials recovery.

Disposal of unprocessed refuse

The two options for disposal are in the oceans (or other large bodies of water) and on land. Disposal in the oceans is, for the most part, illegal in most developed nations. However, ocean disposal is still widely used, either openly or covertly, and it represents a disgrace to standards of public health and environmental engineering. The place for solid waste disposal on land, called a dump, is an inexpensive means of solid waste disposal. The operation of a dump is simple and involves nothing more than making sure that the trucks empty out at the proper spot. Volume is often reduced by setting the refuse on fire, which prolongs dump life.   Rodents, odor, air pollution, and insects at the dump, however, can become serious public health and aesthetic problems, and an alternative method of refuse disposal is necessary.

In the United States, dumps have been rendered obsolete and have been replaced mostly by sanitary land-fills.   A sanitary landfill differs from an open dump in that the latter are simply places to dump wastes but sanitary landfills are engineered operations, designed, and operated according to acceptable standards. Sanitary landfilling involves two principles: compaction of the refuse and placement of a cover. After the refuse is unloaded, it is compacted with heavy machinery, and covered daily with compacted soil.   This cover is from 6 to 12 in thick. Daily cover of the refuse is the single feature that renders a landfill much less of a nuisance and health hazard than is a dump.After closure, a landfill continues to subside, so permanent structures cannot be built on it, but can be used for recreation, parks, or greenbelts.

Landfills

The *sanitary landfill* became the accepted method of disposal because it was considered environmentally sound and reasonably inexpensive. Unfortunately, appropriate land is not always available, and even if it is, landfilling is not always successful. The operation of landfills is becoming increasingly expensive, and there is well-placed concern about throwing away materials that might prove useful. As a result, the idea of processing wastes to reclaim materials and energy was born. Resource recovery is the preferred direction. The proper selection of a landfill site is important. Engineering aspects include:

1. drainage—rapid runoff will lessen mosquito problems, but proximity to streams or well supplies might result in water pollution

2. wind—it is preferable that the landfill be downwind from the community

3. distance from collection

4. size

In years past, sanitary landfills were often indistinguishable from dumps and gained a reputation for being bad neighbors. In recent years as more landfills have been operated properly, it has even been possible to enhance property values with a closed landfill site since such a site must remain open space. Acceptable operation and eventual enhancement of the property are understandably difficult to explain to a community. The landfill operation is actually a biological method of waste treatment. Municipal refuse deposited as a fill is anything but inert. In the absence of oxygen, anaerobic decomposition steadily degrades the organic material to more stable forms. But this process is very slow. After 25 years, the decomposition can still be going strong.

The liquid produced during the decomposition process, as well as the water that has seeped through the groundcover and worked its way out of the refuse, is known as leachate. This liquid, although small in volume, is extremely high in its capacity to pollute. The effect of leachate on groundwater can be severe. The leachate can pollute wells around a landfill to the point where they cease to be a source of potable water. A second by-product of a landfill is gas. Landfills are anaerobic biological reactors and they produce methane and carbon dioxide.

The meaning of the word **_aerobic_** is "living, active, or occurring only in the presence of free oxygen." The meaning of the word _anaerobic_ is "living, active, or occurring only in the absence of free oxygen."

Chapter 5. Soils and Solid Waste

Landfills go through four distinct stages. The **first stage** is aerobic and may last up to several months during which time aerobic organisms are active and affect the decomposition. Once the organisms use up all available oxygen, however, the landfill enters the **second stage** where anaerobic decomposition begins and the acid formers cause a buildup of carbon dioxide (chemical formula CO_2).. The **third stage** is one in which there is a buildup of the anaerobic methane (chemical formula CH_4). The **fourth stage** is the steady-state condition when the fractions of CO_2 and CH_4 are about equal and microbial activity has stabilized.

The biological aspects of landfills as well as the structural properties of compacted refuse limit the ultimate uses of landfills. Uneven settling is often a problem. Landfills should never be disturbed. Not only will this cause additional structural problems, but also trapped gases can be a hazard. A landfill usually represents the least-cost method of acceptable solid waste disposal.

Resource Recovery

It is desirable to have volume reduction prior to waste disposal. Refuse is a bulky material that does not compact easily and thus the volume requirements in landfills are significant. Where land is expensive, the costs of landfilling can be high. Accordingly, various methods of reducing the volume of refuse to be disposed of have been found to be effective. Under the right circumstances, incineration can be an effective treatment of municipal solid waste. Incineration can reduce the volume of waste by a factor of 10 or 20, and the incinerator ash is usually more stable than is the municipal solid waste (MSW) itself. Disposal of the ash can be problematic since heavy metals and some toxic materials will be concentrated in the ash. Incinerators have high capital costs and operating expenses. Air pollution control has effectively doubled the cost of incineration.

Shredding solid wastes (also known as pulverizing) and then spreading the material on fields has been successful in a number of places. The organics are ground up and are of no interest to rats, and spreading dries the refuse thus avoiding odor and fly problems.

The shredded material does not have to be covered with dirt—a significant advantage over the landfill.

At one time, it was easy to dispose of solid waste by dumping it somewhere out of sight. However, this is no longer possible, and it is becoming increasingly difficult to get rid of the waste so it no longer impacts society. One potential solution would be to simply redefine solid waste as a resource, and use it for the production of goods for people. In fact, it is becoming increasingly difficult to find new sources of energy and materials to feed our industrial society. Concurrently, we are finding it more and more difficult to locate solid waste disposal sites, and mainly because of transportation requirements, the cost of disposal is escalating. These two factors—energy and material shortages and fewer and more expensive disposal options—have fueled a new technology called resource recovery.

Except for the process of mass burning of raw refuse, for energy production, the recovery of resources from refuse is primarily a quest for purity. Pure materials can be obtained from mixed municipal solid waste in one of two ways:

> 1. Separation of the materials is performed by the user, the person who decides to discard the various consumer products. This is commonly known as *recycling.*
>
> 2. Separation is performed after the mixed refuse is collected at a central processing facility. This is commonly referred to as *recovery.*

There are only two incentives that could be used to convince the public to undertake the separation of refuse components. The first is regulatory, where the governmental agency dictates that only separated material will be picked up. The second means of achieving cooperation in recycling programs is to appeal to the sense of community spirit and the ethics of environmental concern. Unfortunately, the participation in a recycling project is much lower.

Chapter 5. Soils and Solid Waste

There is a wide gap between what people say they will do and how they actually perform.

As difficult as recycling is and as small as the fraction of materials removed from the waste stream might be, recycling is still a success in many communities. A number of communities are now achieving 20 percent diversion—mostly newspapers, aluminum cans, and clear glass. The EPA has recently established a national goal of 25 percent recycling, and it appears that many communities will meet this objective. The savings in disposal costs are so significant that several large cities have set much higher, and often unrealistic, goals.

Improvements in separated material collection vehicle design over the past few years have been significant. The most effective system seems to be the inclusion of a work station on the truck where the collector can sort the material from the recycling containers while the driver proceeds to the next stop. In this way, undesirable materials can be separated and the product is assured of high purity. Finally, it is absolutely necessary to establish a sense of community mission or cooperation if a recycling program is to succeed.

Most processes for separation of the various materials in refuse rely on a characteristic or property of the specific material as a code, and this code is used to separate the material from the rest of the mixed refuse. Before such separation can be achieved, however, the material must be in separate and discrete pieces, a condition clearly not met by most components of mixed refuse. A common tin can, for example, contains steel in its body, zinc on the seam, a paper wrapper on the outside, and perhaps an aluminum top. Other common items in refuse provide equally or even more challenging problems in separation. One means of assisting in the separation process is to decrease the particle size of refuse, thus increasing the number of particles and achieving a greater number of clean particles. This size reduction step, although not strictly materials separation, is commonly a first step in a solid waste processing facility.

Commonly called shredding, the process of size reduction usually consists of a brute-force breakage of particles by swinging hammers in an enclosure. The shredded refuse is often run over screens that separate materials solely by size and do not identify the material by any other property.

> **Air Classifiers.** Materials can be separated according to their aerodynamic properties. In shredded MSW, most of the aerodynamically light materials are organic, and most of the heavy materials are inorganic, thus air classification can produce a refuse-derived fuel (RDF) superior to unclassified shredded refuse.
>
> **Magnets.** Ferrous material is removed using magnets that continually extract the ferrous and reject the remainder. With the belt magnet, recovery of ferrous is enhanced by placing the belt close to the refuse, but this also decreases the purity of the product. A major problem in using belt magnets is the depth of the refuse on the conveyor belt. The heavy ferrous particles tend to settle to the bottom of the refuse carried on a conveyor, and these are then the farthest away from the magnet.

Energy recovery is desirable. The organic fraction of refuse is a useful secondary fuel. This shredded and classified product can be used in existing power plants either as a supplemental fuel with coal or fired as the sole fuel in separate boilers. Energy recovery from unprocessed refuse (known as mass burn) involves large mass burn units that use unprocessed refuse as fuel. The refuse is burned to produce heat that is in turn used to drive turbines for electrical power production. Mass burn plants are about twice as expensive as landfills (even after the sale of the electricity) but are often the only alternative for communities that are unable to site new landfills.

The solid waste problem must be attacked from the source as well as from disposal. Solid wastes, often called the third pollution, are only now being considered a problem equal in magnitude to that of air and water pollution. It has been suggested that one solution to the

Chapter 5. Soils and Solid Waste

solid waste problem is the development of truly biodegradable forms of materials such as plastics and glass. We are still many years away from the development and use of fully recyclable or biodegradable materials.

Hazardous Waste

Chemical wastes have been the necessary by-products of industrialized societies. Historically the choice of disposal sites was made with little or no attention to potential impacts on groundwater quality and runoff to streams and lakes. Disposal problems were solved by simply piling or dumping these waste products out back. Attitudes began to change in the 1960s and 1970s when air, water, and land were no longer viewed as commodities to be polluted without considering the potential problems of cleanup. Individuals responded with court actions against pollution, and governments responded with revised local zoning ordinances, updated public health laws, and new major federal Clean Air and Clean Water Acts. In 1976, the Federal Resource Conservation and Recovery Act (RCRA) was enacted to give the U.S. Environmental Protection Agency specific authority to regulate the generation, transport, and disposal of hazardous waste. Enactment of RCRA has initiated much needed research into methods of detoxifying or stabilizing hazardous wastes. Engineers are still investigating the most effective ways of managing hazardous waste.

Hazardous waste engineering involves tracing the quantities of wastes generated in the nation from handling and processing options through transportation controls to resource recovery and ultimate disposal alternatives. The word ***corrosivity*** refers to the quality of being corrosive.

A hazardous waste is defined by the degree of ignitability, corrosivity, reactivity, and/or toxicity. This definition includes acids, toxic chemicals, explosives, and other harmful or potentially harmful waste. Radioactive wastes are also hazardous, but because their generation, handling, processing, and disposal differ so drastically from nonradioactive hazards, the radioactive waste problem is addressed separately.

Millions of tons of hazardous waste are generated annually. More than 60 percent is generated by chemical and allied products industried, and the machinery, primary metals, paper, and glass products industries each generate between 3 and 10 percent of the total. Approximately 60 percent of the hazardous waste is liquid or sludge. Most waste is disposed of on the generator's property in inadequately designed and operated pits, ponds, landfills, and incinerators. The hazardous waste problem is compounded by two considerations: The wastes are generated and disposed of in areas where it rains and in areas where people rely on aquifers for supplies of drinking water. Most hazardous waste is generated and inadequately disposed of in the eastern portion of the country. In this region, the climate is wet with patterns of rainfall that permit infiltration and/or runoff to occur. Infiltration permits the transport of hazardous waste into groundwater supplies, and surface runoff leads to the contamination of streams and lakes. Also most hazardous waste is generated and disposed of in areas where people rely on aquifers for drinking water. Major aquifers and well withdrawals underlie areas where the wastes are generated.

Hazardous wastes are transported by trucks, rail flatcars, and barges. Transportation of hazardous wastes presents the same hazards and is regulated in the same way as is transportation of other hazardous materials like gasoline. Because many hazardous wastes are often generated in relatively small quantities, truck transportation—often small-truck transportation—is a highly visible and constant threat to public safety and the environment. There are four basic elements in the control strategy for the movement of hazardous waste from a generator.

Waste Processing and Handling

Waste processing and handling are key concerns as a hazardous waste begins its journey from the generator site to a secure, long-term storage facility. Ideally, the waste can be stabilized, detoxified, or somehow rendered harmless in a treatment process similar to those outlined briefly below.

Chemical Stabilization/Fixation. In these processes, chemicals are mixed with waste sludges, the mixture is pumped onto land, and solidification occurs in several days or weeks. The result is a chemical nest that entraps the waste, and pollutants such as heavy metals may be chemically bound in insoluble complexes. Proponents of these processes have argued for building roadways, dams, and bridges using a selected cement as the fixing agent. The environmental adequacy of the processes has not been documented, however, as long-term leaching and de-fixation potentials are not well understood.

Volume Reduction. Volume reduction is usually achieved in an incineration process. This process takes advantage of the large organic fraction of waste being generated by many industries, but may lead to secondary problems for hazardous waste engineers: air emissions in the stack of the incinerator and ash production in the base of the incinerator. Both byproducts of incineration must be addressed in terms of legal, cost, and ethical constraints. Because incineration is often considered a very good method for the ultimate disposal of hazardous waste, we discuss it in some detail later in this chapter.

Waste Segregation. Prior to shipment to a processing or long-term storage facility, wastes are segregated by type and chemical characteristics. Similar wastes are grouped in a 55-gallon drum or group of drums, segregating liquids like acids from solids such as contaminated lab clothing and animal carcasses. Waste segregation is generally practiced to prevent undesirable reactions at disposal sites and may lead to economies of scale in the design of detoxification or resource recovery facilities.

Detoxification. Numerous thermal, chemical, and biological processes are available to detoxify chemical wastes. Options include neutralization, ion exchange, incineration, pyrolysis, aerated lagoons, and waste stabilization ponds. These technologies are extremely waste-specific; ion exchange obviously does not work for every chemical and some forms of heat treatment can be prohibitively expensive for sludges that have a high water content. It is time to call in the chemical engineers whenever detoxification technologies are being considered.

Degradation. Methods exist that chemically degrade some hazardous wastes and render them safer, if not completely safe. Chemical degradation processes, which are very waste-specific, include hydrolysis, to destroy organophosphorus and carbonate pesticides, and chemical dechlorination, to destroy some polychlorinated pesticides. Biological degradation generally involves incorporating the waste into the soil. Landfarming, as it has been termed, relies on healthy soil microorganisms to metabolize the waste components. Such land-farming sites must be strictly controlled for possible water and air pollution that results from overactive or underactive organism populations. For the most part, degradation of hazardous waste is in the research and development stages.

Encapsulation. A wide range of material is available to encapsulate hazardous waste. Ranging from the basic 55-gallon steel drums used throughout the nation, options include concrete, asphalt, and plastics. Several layers of different materials are often recommended such as a steel drum coated with an inch or more of polyurethane foam to prevent rust.

Resource Recovery Alternatives

Resource recovery alternatives are based on the premise that one person's waste is another person's prize. What may be a worthless drum of electroplating sludge to the plating engineer may be a silver mine to an engineer skilled in metals recovery. In hazardous waste management, the waste must be transferred to a location where it is viewed as a resource.   Wastes generally recognized as having transfer value are: (1) wastes having high concentrations of metals,

Chapter 5. Soils and Solid Waste

(2) solvents, (3) concentrated acids, (4) oils, and (5) combustibles for fuel. One person's waste can truly be another person's valued resource.

Siting Considerations must be taken into account. In selecting a site, the following must be considered: hydrology, climatology, geology, and ecology as well as current land use, environmental health, and transportation. Risk analysis is an important part of the siting process.

> **Hydrology**. Hazardous waste landfills should be located well above historically high groundwater tables. Care should be taken to ensure that a location has no surface or subsurface connection such as a crack in confining strata, between the site and a watercourse. Hydrological considerations limit direct discharge of wastes into groundwater or surface water supplies.
>
> **Climatology**. Hazardous waste management facilities should be located outside the paths of recurring severe storms. In addition, areas of high air pollution potential should be avoided in site selection processes so that winds and/or inversions will not act to hold pollutants close to the surface of the earth.
>
> **Geology**. A disposal or processing facility should only be located on stable geologic formations. Impervious rock, which is not littered with cracks and fissures, is an ideal final liner for hazardous waste landfills.
>
> **Ecology**. The ecological balance must be considered as hazardous waste management facilities are located in a region. Ideal sites in this respect include areas of low fauna and flora density, and efforts should be made to avoid wilderness areas, wildlife refuges, and migration routes.
>
> **Alternative Land Use**. Areas with low ultimate land use should receive prime consideration as facilities are sited in a region. Areas with high recreational use potential should be

avoided because of the increased possibility of direct human contact with the wastes.

Environmental Health. Landfills and processing facilities should be located away from private wells, away from municipal water supplies, and away from high population densities. Flood plains should be avoided, at least up to the 100-year storm level.

Transportation. Transportation routes to facilities are a major consideration in siting hazardous waste management facilities. Such facilities should be accessible by all-weather highways to avoid spills and accidents during periods of rain and snowfall. Ideally, the closer a facility is to the generators of the waste, the less likely are spills and accidents as the wastes move along the countryside.

Socioeconomic Factors. Factors that could make or break an effort to site a hazardous waste management facility fall under this major heading. Such factors range from citizen acceptance to long-term care and monitoring of the facility.

Mixed Waste. The term mixed waste refers to mixtures of hazardous and radioactive wastes. (Organic solvents used in liquid scintillation counting are an excellent example.) Siting a new mixed waste facility is virtually impossible at present because in many cases the laws governing handling of chemically hazardous waste conflict with those governing handling of radioactive waste. Existing radioactive waste disposal sites contain some mixed waste.

Incinerators. Incineration is a controlled process that uses combustion to convert a waste to a less bulky, less toxic, or less noxious material. The principal products of incineration from a volume standpoint are carbon dioxide, water, and ash, but the products of primary concern due to their environmental effects are compounds containing sulfur, nitrogen, and halogens. When the gaseous combustion

Chapter 5. Soils and Solid Waste 121

products from an incineration process contain undesirable compounds, a secondary treatment such as afterburning, scrubbing, or filtration is required to lower concentrations to acceptable levels prior to atmospheric release. The solid ash products from the incineration process are a major concern and must reach adequate ultimate disposal.

The **advantages** of incineration as a means of disposal for hazardous waste are

1. Burning wastes and fuels in a controlled manner has been carried on for many years and the basic process technology is available and reasonably well developed. This is not the case for some of the more exotic chemical degradation processes.

2. Incineration is broadly applicable to most organic wastes and can be scaled to handle large volumes of liquid waste.

3. Large expensive land areas are not required.

The **disadvantages** of incineration include:

1. The equipment tends to be more costly to operate than many other alternatives.

2. It is not always a means of ultimate disposal in that normally an ash remains that may or may not be toxic but that in any case must be disposed of properly.

3. Unless controlled by applications of air pollution control technology, the gaseous and particulate products of combustion can be hazardous to health or damaging to property.

The decision to incinerate a specific waste will therefore depend first on the environmental adequacy of incineration as compared to other alternatives and second on the relative costs of incineration and the environmentally sound alternatives.

The type and form of waste will dictate the type of combustion unit required. A number of control methods have been successfully developed for applications where the pollutants are in the form of fumes or gas. If the waste gas contains organic materials that are combustible, incineration should be considered as a final method of disposal. When the amount of combustible material in the mixture is below the lower flammable limit, it may be necessary to add small quantities of natural gas or other auxiliary fuel to sustain combustion in the burner. Thus economic considerations are critical in the selection of incinerator systems because of the high costs of these additional fuels.

Boilers for some high-temperature industrial processes can serve as incinerators for toxic or hazardous carbonaceous waste. Cement kilns, which must operate at temperatures in excess of 1400°C (2550°F) in order to produce cement clinker, can use organic solvents as fuel, and this provides an acceptable method of waste solvent and waste oil disposal.

Incineration is also a possibility for the destruction of liquid wastes. Liquid wastes may be classified into two types from a combustion standpoint: combustible liquids and partially combustible liquids. Combustible liquids include all materials having sufficient caloric value to support combustion in a conventional combustor or burner. Noncombustible liquids cannot be treated or disposed of by incineration and include materials that would not support combustion without the addition of auxiliary fuel and would have a high percentage of noncombustible constituents such as water.

When starting with a waste in liquid form, it is necessary to supply sufficient heat for vaporization in addition to raising it to its ignition temperature. In order that a waste may be considered combustible, several rules of thumb should be used. The waste should be pumpable at ambient temperature or capable of being pumped after heating to some reasonable temperature level. Since liquids vaporize and react more rapidly when finely divided in the form of a spray, atomizing nozzles are usually employed to inject waste liquids into incineration equipment whenever the viscosity of the waste permits

Chapter 5. Soils and Solid Waste

atomization. If the waste cannot be pumped or atomized, it cannot be burned as a liquid but must be handled as sludge or solid.

Incineration is not a total disposal method for many solids and sludges because most of these materials contain non-combustibles and have residual ash. Complications develop with the wide variety of materials that must be burned. Controlling the proper amount of air to give combustion of both solids and sludge is difficult, and with most currently available incinerator designs, this is impossible. The types of incinerators applicable to solid wastes are open pit incinerators and closed incinerators such as rotary kilns and multiple hearth incinerators. Generally, the incinerator design does not have to be limited to a single combustible or partially combustible waste. Often it is both economical and feasible to utilize a combustible waste, either liquid or gas, as the heat source for the incineration of a partially combustible waste that may be either liquid or gas.

Experience indicates that wastes, which contain only carbon, hydrogen, and oxygen and which can be handled in power generation systems, can be destroyed in a way that reclaims some of their energy content. These types of wastes may also be judiciously blended with wastes having low energy content such as the highly chlorinated organics in order to minimize the use of purchased fossil fuel. On the other hand, rising energy costs will not be a significant deterrent to the use of thermal destruction methods where they are clearly indicated to be the most desirable method on an environmental basis.

Air emissions from hazardous waste incinerators include the common air pollutants. In addition, inadequate incineration can result in emission of some of the hazardous materials that the incineration was intended to destroy. Incomplete combustion, particularly at relatively low temperatures, can also result in production of a class of compounds known collectively as *dioxins,* including both polychlorinated dibenzodioxins (PCDD) and polychlorinated dibenzofurans (PCDF). The compound in this class that has been identified as a carcinogen and teratogen is 2, 3, 7, 8-

tetrachlorodibenzo-p-dioxin (2,3,7,8-TCDD). There is growing public concern about TCDD emissions from incinerators.

TCDD was first recognized as an oxidation product of trichlorophenol herbicides (2, 4-D and 2, 4, 5-T, one of the ingredients of Agent Orange). In 1977, it was one of the PCDDs found present in municipal incinerator fly ash and air emissions and has subsequently been found to be a constituent of gaseous emissions from virtually all combustion processes, including trash fires and barbecues. TCDD is degraded by sunlight in the presence of water.

Teratogenesis refers to the process by which congenital malformations are produced in an embryo or fetus. The acute toxicity of TCDD in animals is extremely high (LD50 in hamsters of 3.0 mg/kg; carcinogenesis and genetic effects (teratogenesis) have also been observed in chronic exposure to high doses in experimental animals. In humans, the evidence for these adverse effects is mixed. While acute effects like skin rashes and digestive difficulties have been observed on high accidental exposure, these are transitory. Public concern has focused on chronic effects, but existing evidence for either carcinogenesis or birth defects in humans from chronic TCDD exposure is inconsistent. Regulations governing incineration are designed to limit TCDD emission to below measurable quantities; these limits can usually be achieved by the proper combination of temperature and residence time. Engineers should understand, however, that public concern about TCDD (and dioxins in general) has occasionally reached hysterical proportions and is a major factor in opposition to incinerator siting.

Landfills must be adequately designed and operated if public health and the environment are to be protected. The general components that go into the design of these facilities, as well as the correct procedure to follow during the operation and post-closure phase of the facility's life, are discussed below.

Design

Chapter 5. Soils and Solid Waste

Three levels of safeguard must be incorporated into the design of a hazardous landfill. These levels are displayed in Figure 14-3. The primary system is an impermeable liner, either clay or synthetic material, coupled with a leachate collection and treatment system. Infiltration can be minimized with a cap of impervious material overlaying the landfill and sloped to permit adequate runoff and to discourage pooling of the water.

The objectives are to prevent rainwater and snow melt from entering the soil and percolating to the waste containers and, in case water does enter the disposal cells, to collect and treat it as quickly as possible. Side slopes of the landfill should be a maximum of 3:1 to reduce stress on the liner material. Research and testing of the range of synthetic liners must be viewed with respect to a liner's strength, compatibility with wastes, costs, and life expectancy. Rubber, asphalt, concrete, and a variety of plastics are available, and such combinations as polyvinyl chloride overlaying clay may prove useful on a site-specific basis.

A leachate collection system must be designed by contours to promote movement of the waste to pumps for extraction to the surface and subsequent treatment. Plastic pipes, or sand and gravel, similar to systems in municipal landfills and used on golf courses around the country, are adequate to channel the leachate to a pumping station below the landfill. One or more pumps direct the collected leachate to the surface where a wide range of waste-specific treatment technologies are available, including:

1. **Sorbent material**: carbon and fly ash arranged in a column through which the leachate is passed

2. **Packaged physical-chemical units**, including chemical addition and flash mixing, controlled flocculation, sedimentation, pressure filtration, pH adjustment, and reverse osmosis

The effectiveness of each method is highly waste specific and tests must be conducted on a site-by-site basis before a reliable leachate treatment system can be designed. All methods produce waste sludges that must reach ultimate disposal.

A secondary safeguard system consists of another barrier contoured to provide a backup leachate collection system. In the event of failure of the primary system, the secondary collection system conveys the leachate to a pumping station, which in turn relays the wastewater to the surface for treatment.

A final safeguard system is also advisable. This system consists of a series of discharge wells up-gradient and down-gradient to monitor groundwater quality in the area and to control leachate plumes if the primary and secondary systems fail. Up-gradient wells act to define the background levels of selected chemicals in the groundwater and to serve as a basis for comparing the concentration of these chemicals in the discharge from the down-gradient wells. This system thus provides an alarm mechanism if the primary and secondary systems fail. If gas generation is possible in a hazardous waste landfill, a gas collection system must be designed into the landfill. Sufficient vent points must be allowed so that the gas generated may be burned off continuously or processed prior to its emission into the atmosphere.

Operation

As waste containers are brought to a landfill site for burial, specific precautions should be taken to ensure the protection of public health, worker safety, and the environment. Wastes should be segregated by physical and chemical characteristics and buried in the same cells of the landfill. Three-dimensional mapping of the site is useful for future mining of these cells for resource recovery purposes. Observation wells with continuous monitoring should be maintained, and regular core soil samples should be taken around the perimeter of the site to verify the integrity of the liner materials.

Site Closure

Once a site is closed and does not accept more waste, the operation and maintenance of the site must continue. The impervious cap on top of the landfill must be inspected and maintained to minimize infiltration. Surface water runoff must be managed, collected, and possibly treated. Continuous monitoring of surface water, groundwater, soil, and air quality is necessary as ballooning and rupture of the cover material may occur if gases produced and/or released from the waste rise to the surface. Waste inventories and burial maps must be maintained for future land use and waste reclamation. A major component of postclosure management is maintaining limited access to the area.

Hazardous waste is a relatively new concern of environmental engineers. For years, the necessary by-products of an industrialized society were piled out back on land that had little value. As time passed and the rains came and went, the migration of harmful chemicals moved hazardous waste to the front page of the newspaper and into the classroom. Engineers employed in all public and private sectors must now face head on the processing, transport, and disposal of these wastes. Hazardous waste is appropriately addressed at the front end of the generation process: either in maximizing resource recovery in house, using industry-wide clearinghouses and exchanges, or detoxification at the site of generation. Storage, landfilling in particular, is at best a stopgap measure for hazardous waste handling.

Chapter 6. Water

"Smooth runs the water where the brook is deep."—Shakespeare (*Henry VI, Part II*)

Hydrologic Cycle

About 70 percent of the earth is covered in water! The water falls into five categories: oceans, rivers, lakes, groundwater, and ice. As an essential part of our daily diet, water is vital for the survival of our ecosystem. The oceans— Atlantic, Pacific, Indian, Arctic, and Southern —comprise 97.3% of the total water on earth. The volume of the ocean is in the range of 1.3 to 1.5 billion cubic kilometers. Additional water is bound up in hydrated minerals. All the various sources of water on the Earth are together called the hydrosphere and the study of water is called hydrogeology. We are still using the oceans as a big sink. Is this sink so large that it will never be ruined? No, it is already polluted; international regulation is necessary to assure its healthy survival. This pollution can affect the marine life cycle, and if this cycle is disrupted it will also affect life cycles on land, including that of human beings. As seen in photographs from space, the Earth is a dramatic blue planet, the only planet in the Solar system that contains abundant liquid water. Water is a major feature of the surface of the Earth.

Water is critical to many near-surface processes (including weather). Water has unique chemical and physical characteristics and it is important as a solvent. An increasing quantity of water is needed for drinking, industrial use, and irrigation. Water is obtained from surface and subsurface sources. The basics of water distribution and flow (both at the surface and in the subsurface) are vital in environmental studies. Although natural factors may degrade water quality, it is manmade pollutants that are becoming increasingly important. The nature of the pollutants, their sources, and detection is described in this chapter.

Water, a familiar substance, has unusual properties. The formula for water, H_2O, is deceptively simple and covers up many of its

Chapter 6. Water

complexities. The bonding in a water molecule is covalent. The oxygen is 2 electrons short of a complete outer shell of 8 electrons, and it gets them by sharing with a couple of hydrogens. The four pairs of electrons are distributed in a tetrahedral arrangement with the 2 bonding pairs attached to hydrogens, and because of this the H-O-H molecule is bent at 104.5°. The charges in the water molecule are distributed so that there is a slight positive excess on the hydrogens and a slight excess negative on the oxygens. This is part of the reason why the molecules are attracted to one another, making water a liquid at room temperature. It is also part of the reason why water is a good solvent for ionic compounds. Liquid water is also unusual in having its maximum density at about 4°C and in having its solid form (ice) less dense than liquid water. We are so accustomed to seeing ice float that it is easy to overlook how unusual this behavior is. Virtually all other solids sink in their liquids.

The covalent water molecules have a slight tendency to form ions. This self-ionization can be written:

$$H_2O = H^+ + OH^-$$

or more realistically as:

$$2H_2O = H_3O^+ + OH^-$$

In pure water the concentration of H^+ (strictly H_3O^+) has to be equal to the concentration of OH^- and both are 10^{-7} moles/liter. The hydrogen ions (H^+) are responsible for acidity and are usually expressed as a pH scale, where pH is the negative power of the hydrogen ion concentration. In a neutral solution this concentration is 10^{-7} moles/liter and so the pH is 7. As hydrogen ion concentration increases the solution becomes more acidic and the pH value decreases. Conversely in a basic, or alkaline, solution the pH value will be greater than 7.

A person needs about 3 liters of water per day for survival. However, an average American uses 200 to 300 liters per day for personal use. A single toilet flush uses approximately 20 liters. With the inclusion of industrial use the average rises to 5450 liters per day of fresh water per capita. Most uses require relatively fresh water for drinking, agriculture, and many industrial processes. Water is conserved and is not used up, but getting fresh water takes considerable energy (for example, in desalinating sea water). Water is continuously cycled among the different environments. These various near-surface locations for water are interrelated by the "hydrologic cycle" which links water evaporation, condensation, precipitation, and flow back to the oceans, either directly or through groundwater.

Water evaporates from the surface of all bodies of water including the oceans, lakes and rivers, but is quantitatively most important for the oceans simply because the surface is so much greater. Evaporation is favored by higher temperatures, lower humidity, higher wind speed, and fresher waters. Loss of water from the land surface is estimated by measuring the loss from standard pans, most commonly the National Weather Service Class A pan which is 48 in. in diameter and 10 in. deep. The relationship between evaporation in the limited area of the pan and the real world was established by the Lake Heffner Experiment. The 2300 acre Lake Heffner in Oklahoma City was heavily instrumented and monitored for 15 months. Evaporative losses can be an important factor in water resource projects and depending on flow conditions, transport distance, and temperature may be considerable. In one Colorado project evaporative loss from an irrigation canal network ran as high as 28 percent during the 12 day time of transport.

Water vapor is an invisible gas, but as temperature falls it condenses back to a liquid and forms visible droplets. This may occur as spectacular cumulus clouds, or as the condensation on a cold bottle of beer. In the atmosphere condensation usually occurs on nuclei that may be dust particles, salt crystals, smoke particles, etc. Droplets in clouds average about 0.01 mm in diameter, but generally do not fall as precipitation until they have coalesced into larger droplets.

Chapter 6. Water

Water vapor in the atmosphere condenses and falls back to earth as rain, snow, or hail. Even in the centers of continents most of the water vapor is derived by evaporation from the oceans, but rainfall amounts are locally very variable (see Figure). Winds may force moisture-laden air up over mountain ranges and as it rises it cools, leading to condensation and precipitation. This is called "orographic precipitation" and the best example in the U.S. is provided by the mountains of the Pacific coast which have considerable rain and snow, while in the so-called "rain shadow" to the east precipitation is very much less. The lowest average rainfall in the U.S. is in the rain shadow at Death Valley where less than 2 inches per year falls. A similar situation produces precipitation when warm air is forced up over cold air in a "warm front."

Infiltration describes the flow of water from the land surface down into the soil and is an important part of the hydrologic cycle. This water is vitally important to the growth of plants and eventually leads to the recharge of subsurface aquifers. The soil zone may be unsaturated with the pore spaces only partially filled with water, but at some depth the pores become completely filled and this is the "water table" that marks the top of the ground water. The rate of infiltration depends on soil permeability, the minerals making up the soil, and temperature.

Three major processes act to control the return of moisture to the atmosphere over vegetated land areas. They are:

    1. Evaporation of precipitation intercepted by plant surfaces

    2. Evaporation of moisture from plants through transpiration

    3. Evaporation of moisture from the soil surface

Evapotranspiration is responsible for returning roughly two-thirds of the annual precipitation back to the atmosphere in temperate zones, but can rise to greater than 90 percent in arid areas. Plants are classified by the way in which they use water:

1. Hydrophytes: Grow in water or have their roots in water all the time

2. Mesophytes: Land plants needing average amounts of water

3. Xerophytes: Plants that live in arid climates

4. Phreatophytes: Plants that send their roots very deep to reach the water table.

Water in the subsurface, or groundwater, is probably the least familiar part of the hydrologic cycle. However, it plays a significant role, and water that penetrates deep into the subsurface may take thousands of years before it rejoins the rest of the hydrologic cycle. Even in prehistoric times wells drilled in arid areas proved the presence of large quantities of subsurface water, and groundwater is still important, even in temperate regions, in supplying water for drinking and agriculture.

The final step in completing the hydrologic cycle is runoff which returns water to the oceans. Here streams and rivers play an important role and flow in open channels forms an important study area. Rivers drain their respective "drainage basins", and all of the water that falls as rain in that basin (except evaporative losses) eventually flows to the sea. In arid areas rain may flow quickly across the surface and river flow increases rapidly after a storm producing a "flash flood". In vegetated areas the water is held up longer and there is more time for infiltration so that the peak water flow in the streams and rivers is spread over a longer time. The flow in rivers is measured by a hydrograph.

Chemical composition

The composition of waters near the Earth's surface varies widely, and is an important factor that controls their usefulness. Natural waters always contain some dissolved constituents, and they frequently also have solids in suspension. The total amount of dissolved material is conveniently given as "Total Dissolved Solids", or TDS, and is expressed as grams of dissolved material in a kilogram

Chapter 6. Water

of solution. This is close to "parts per thousand" and salinities are often given in this way, as well as in "parts per million" (ppm). Fresh water has less than 1000 ppm (i.e. 1 ppt) and categories of fresh and salty water suggested by the U.S.G.S. are shown in their publications. The U.S. Public Health Service Drinking Water Standards recommends water with a maximum of 500 ppm TDS (i.e. 0.5 ppt) for drinking.

SEA WATER. Sea water has a salinity of 35,000 ppm with sodium and chloride being the most abundant ions. pH is close to 8.0.

LAKES. Compositions vary widely from fresh to highly saline (e.g. Great Salt Lake and the Dead Sea). It is not uncommon for lakes to stratify with less oxygenated (and often more saline) waters at the bottom.

RIVERS. Compositions vary but are generally dominated by calcium and bicarbonate ions.

OPEN CHANNEL FLOW. Water flows across the land surface in streams and rivers, and in a variety of man-made culverts and aqueducts. It flows in a less confined way as sheet flows across hill sides and parking lots. Water flowing in a channel is in contact with the bottom and sides as well as the atmosphere. These have different frictional effects so that water flows more slowly near the surfaces and faster in the center. Flow rate is often estimated using the Manning formula which was introduced by an Irish engineer in 1889. In the United States, in practice, it is very frequently called simply Manning's equation. The Manning formula is an empirical formula estimating the average velocity of a liquid flowing in a conduit that does not completely enclose the liquid, i.e., open channel flow. However, this equation is also used for calculation of flow variables in case of flow in partially full conduits, as they also possess a free surface like that of open channel flow. All flow in so-called open channels is driven by gravity.

The **Manning Equation** is a discharge formula. It gives volumetric water flow rate Q as a function of channel roughness n, cross-sectional area A, hydraulic radius R, and slope S. The Manning Equation for U.S. units is:

$$Q = \left(\frac{1.49}{n}\right) A \left(R^{\frac{2}{3}}\right)\left(S^{\frac{1}{2}}\right)$$

and for S.I. units is:

$$Q = \left(\frac{1.0}{n}\right) A \left(R^{\frac{2}{3}}\right)\left(S^{\frac{1}{2}}\right)$$

where

Q = volumetric water **flow rate** passing through the stretch of channel, ft³/sec (m³/s for S.I.)

A = cross-sectional **area** of flow perpendicular to the flow direction, in ft² (m² for S.I.)

S = bottom **slope** of channel, in ft/ft (m/m for S.I.), dimensionless

n = Manning **roughness** coefficient (empirical constant), dimensionless

P = wetted **perimeter** of cross-sectional flow area, in ft (m for S.I.)

R = hydraulic **radius** = A/P, in ft (m for S.I.)

$V = Q/A$ = cross-sectional average **velocity**, in ft/second (m/second for S.I.)

Slope provides the driving force, and as slope increases so does water velocity but a slope that is 4 times as steep only increases the velocity by a factor of 2 (i.e. $\sqrt{4}$) The hydraulic radius is the ratio of cross sectional area of the channel to the wetted perimeter. Manning's n is a measure of channel roughness, and rougher channels have higher values of n which lead to a decreased water velocity. Roughness n increases as the channel bottom becomes rockier and as the banks become covered with vegetation. Sharp curves, or meanders, in the channel also increase n.

Chapter 6. Water

At higher velocities water can carry a bigger sediment load. This means that at points where the water in a river or stream slows down it cannot carry as much sediment in suspension and some of the load is deposited. Rivers represent a balance between sediment transport and sediment deposition.

FLOODING BY RIVERS AND STREAMS. Floods are an inevitable consequence of runoff. Nearly all uncontrolled streams regularly overflow their banks and attempts to control flooding by the construction of levees and dikes go back into prehistory. Dams have been built to control flooding (and may also generate hydroelectric power), but they have a limited lifetime (generally about 100 years) and eventually they silt up. Many modern construction projects are very dramatic, such as Glen Canyon dam/Lake Powell which took 20 years to fill with water. The dramatic floods of 1993 along the Mississippi River show how difficult it is to control major river systems.

WATER IN THE GROUND. Water that infiltrates down into the ground becomes part of the groundwater. In most areas the soil near the surface has water adsorbed on the mineral grains, but the pore spaces are full of air. This is the "vadose zone." Deeper in the section and below the top of the water table the pores are completely filled with water and no air is present. Between these two zones is a variable zone called the "capillary zone," where water soaks upwards. The vertical extent of this zone depends on the pore size distribution and the structure of the soil or rock. The "phreatic zone" occurs below the capillary zone and is the top of the groundwater.

Drinking-Water Pollution and Human Health

The quest for pure drinking water in the face of contamination has existed since ancient times. Public water supplies often are a source of infection for humans. Hazardous substances (such as those related to industrialization and agricultural activity) contaminate surface water and ground water. The threat to health posed by water contamination is high. Much of the population use groundwater from wells or springs as their primary source of drinking water and surface water remain the world's main source of drinking water.

Contamination can be difficult to detect, and extremely difficult to reverse once detected. Furthermore, many inorganic and organic chemicals that had been deemed safe are now thought to be hazardous to human health. These problems are being felt almost everywhere.

There are four general characteristics of water quality: (physical, chemical, biologic, and radioactive). Most physical aspects of water quality (taste, odor, color, temperature, turbidity, suspended solids, and mineral content) are not of primary concern with respect to human health. Improved recognition and prevention of contamination by biological agents is of continuing concern, but is well reviewed elsewhere.

Surface water can be contaminated by point or non-point sources. A runoff pipe from an industrial plant or a sewage-treatment plant discharging chemicals into a river is a point source; the carrying of pesticides by rainwater from a field into a lake is a non-point discharge. Fresh surface water can also be affected by groundwater quality; for example, approximately 30% of the streamflow of the United States is supplied by groundwater emerging as natural springs or other seepage. People can be exposed to polluted groundwater or surface water through a number of routes. Too often, contaminated water is collected or pumped and used directly for drinking or cooking.

Groundwater is contained in a geological layer termed an *aquifer*. Aquifers are composed of permeable or porous geological material, and may either be unconfined (and thereby most susceptible to contamination) or confined by relatively impermeable material called **aquitards**. Though they are located at greater depths and are protected to a degree, confined aquifers can nevertheless be contaminated when they are tapped for use or are in proximity for a prolonged period of time to a source of heavy contamination.

Contamination of aquifers can occur via the leaching of chemicals from the soil, from industrial discharges, from septic tanks, or from underground storage tanks. Fertilizers applied to agricultural lands

Chapter 6. Water

contain nitrates which dissolve easily in water. Rainwater percolating through the soil can carry the dissolved nitrates with it into aquifers. Industry or homes can discharge wastewater directly into groundwater from septic tanks or waste holding tanks. Buried underground tanks used to store chemicals, such as gasoline or fuel oil, can leak, allowing their contents to seep into groundwater.

The chemical characteristics of a contaminant may change as it percolates through the soil zone to the aquifer. Attenuation may occur through a number of processes, such as dilution, volatilization, mechanical filtration, precipitation, buffering, neutralization, microbial metabolism, and plant uptake. These generally reduce the toxicity of the contaminant. Once a contaminant gains entry into an aquifer, transport usually results in an elliptical plume of contamination, the shape, flow rate, and dispersion of which depend on aquifer permeability, hydraulic gradients, contaminant chemistry, and many other factors.

Significant exposure to chemicals in surface water can also occur when swimming in a lake or river. Some chemicals accumulate in fish that are subsequently caught and eaten. A chemical that volatilizes easily can escape from groundwater and spread out over a large volume of the soil, and in gaseous form the chemical can then be released into surroundings or can enter homes through cracks in basements, exposing residents through inhalation. If water used for bathing is contaminated, some chemicals can also be absorbed through skin or inhaled in the fine spray of a shower. Of these routes of exposure, use of contaminated water for drinking and cooking is clearly the most dominant threat, followed by ingestion of contaminated fish (especially in areas where high fish consumption and pollution coexist).

Most of the contaminants in surface water and groundwater that are due to human activity derive from agricultural and industrial sources. The spectrum of contaminants is enormous. The most important ones are toxic heavy metals (such as lead, arsenic, cadmium, and mercury), pesticides and other agricultural chemicals (such as nitrates, chlorinated organic chemicals (DDT), organophosphate or carbamate (aldicarb) insecticides, and herbicides, and volatile organic chemicals (such as gasoline products and the halogenated solvents trichloroethene and tetrachloroethene). There are also some natural sources of hazardous chemical exposure; for example, deep wells are often contaminated with naturally occurring arsenic.

Specific Hazards

Nitrates and Nitrites

What is the difference between nitrates and nitrites? Each consists of a single nitrogen atom that is bonded to oxygen atoms. A nitrate NO_3 has three oxygen atoms, while a nitrite NO_2 has two oxygen atoms. Nitrates are harmless, but when they convert to nitrites it may not be good. When nitrates hit the tongue, the bacteria in the mouth or the enzymes in the body turn the nitrates into nitrites. Nitrites can be good when they form nitric oxide, but when they form nitrosamines, it can have a negative effect, even producing cancer-causing cells. When certain foods are eaten, the nitrites are good because they prevent certain harmful bacteria from forming. For example, nitrites are used to cure meat so that it stays pink or red. Otherwise it would turn brown, so no one would want not purchase it. Nitrates naturally occur in fruits, vegetables and grains, and this natural occurrence prevents the formation of harmful nitrosamines. A nitrosamine is a compound containing the group NNO attached to two organic groups. Compounds of this kind are generally carcinogenic. Nitrosamines also develop when nitrites end up the acidic stomach. High temperature and frying both increase the potential for nitrosamines.

From a global perspective, biological processes such as nitrogen fixation and the conversion of organic nitrogen to ammonia (NH_3) or nitrate (NO_3) are the major sources of inorganic nitrogen compounds in the environment. However, on a local scale, municipal and industrial wastewaters (particularly sewage treatment plants, fertilizers, refuse dumps, septic tanks, and other sources of organic waste) are major nitrogen sources. Waste sources are significant as

Chapter 6. Water

they can greatly exceed natural sources, and are increasingly found in groundwater, primarily because of a marked rise in the use of nitrogenous fertilizers around the world.

Most human intake of nitrogen normally comes from food rather than water. Vegetables are the highest source of nitrates for most people. Of lesser importance is the contribution of nitrates from cured meats. Cured meats, baked goods, and cereals contribute most of the dietary nitrite; vegetables provide much less. The total nitrogen content of water is usually measured for both nitrates and nitrites. Nitrate, nitrite, or both are more likely to be found in higher concentrations in groundwater than in surface water, and shallow wells (especially dug wells) are more likely to be contaminated than deep or drilled wells. The drinking-water standard for nitrate in the United States is 10 mg per liter of nitrate (measured as total nitrogen). This level is often exceeded; for example, in well water surveys in South Dakota, up to 39% of dug or bored wells were found to contain nitrates above this level.

Methemoglobinemia is a condition caused by elevated levels of methemoglobin in the blood. Methemoglobin is a form of hemoglobin that contains the ferric form of iron. The affinity for oxygen of ferric iron is impaired. The binding of oxygen to methemoglobin results in an increased affinity for oxygen in the remaining heme sites that are in ferrous state within the same tetrameric hemoglobin unit. This leads to an overall reduced ability of the red blood cell to release oxygen to tissues. When methemoglobin concentration is elevated in red blood cells, tissue hypoxia may occur.

Two potential health effects of concern from nitrate or nitrite in drinking water are the induction of methemoglobinemia. Nitrate itself is relatively non-toxic to humans; however, when converted in the body to nitrite and absorbed, this nitrogen compound is capable of oxidizing hemoglobin (the principal oxygen-transport molecule of the body), with the consequent induction of methemoglobinemia and oxygen starvation. With drinking water contaminated by nitrates at levels above 10 mg/liter, the risk is significant; around

17-20% of infants develop methemoglobinemia when exposed to these higher levels.

Heavy Metals

The heavy metals of greatest concern for health with regard to environmental exposure through drinking water are lead and arsenic. Cadmium, mercury, and other metals are also of concern, although exposure to them tends to be more sporadic. Significant levels of these metals may arise in drinking water, directly or indirectly, from human activity. Of most importance is seepage into groundwater of the runoffs from mining, milling, and smelting operations, which concentrate metals in ores from the earth's crust, and effluents and hazardous wastes from industries that use metals. Lead contamination in dnnking water is of particular concern, as lead was used in household plumbing and in the solder used to connect it. Seepage of heavy metals (especially arsenic) from the earth's crust can be a natural source of contamination in some areas where deep wells are used for drinking water. The accumulation of metals in water, in soil, and in the food chain is accelerating around the world.

Lead

The sources of lead in drinking water that are of greatest concern are lead pipes, the use of which was highly prevalent until the 1940s, and lead solder, which was used (and is still being used) to connect plumbing. Also of concern is the seepage of lead from soil contaminated with the fallout from combusted leaded gasoline and the potential seepage of lead in hazardous-waste sites. Lead contamination of drinking water from lead pipes and solder is more likely to be found in water samples taken at the tap than at the treatment plant. Soft water leaches more lead than does hard water, and the greatest exposure is likely to occur when a tap is first turned on after not being used for six or more hours.

The Environmental Protection Agency's (EPA's) action level for lead in drinking water is 0.015 mg/L (milligrams per liter) or 15 ppb (parts per billion). This level is used to determine when lead problems should be fixed. However, there is no "safe" level of lead in drinking water, because there is no safe level of lead in the body. Lake and river water, worldwide, contains about 1-10 micrograms of lead per liter. Because of lead in plumbing systems, lead levels in drinking water at the tap can be much too high. Drinking water is only one of many potential sources of lead exposure; lead paint, dust, food, and air pollution are other important sources, particularly in old urban areas. CDC now uses a blood lead reference value of 5 micrograms per deciliter to identify children with blood lead levels that are much higher than most children's levels. Evidence links low-level lead exposure to adverse effects on neurobehavioral development and school performance in children. The importance of reducing children's exposure to lead is underscored by new evidence that suggests that cognitive deficits caused by lead are at least partly reversible.

Arsenic

Drinking water is at risk for contamination by arsenic from a number of human activities, including the leaching of inorganic arsenic compounds used in pesticide sprays, the contamination of surface water by fallout from the combustion of arsenic-containing fossil fuel, and the leaching of mine tailings and smelter runoff. With chronic exposure at high levels, children are particularly at risk; the primary symptoms are abnormal skin pigmentation, hyperkeratosis, chronic nasal congestion, abdominal pain, and various cardiovascular manifestations.

At lower levels of exposure, cancer is the outcome of primary concern. Occupational and population studies have linked chronic high-dose arsenic exposure to cancer of the skin, the lungs, the lymph glands, the bone marrow, the bladder, the kidneys, the prostate, and the liver. Using a linear dose-response model to extrapolate risk, and imposing that risk on a large population, one would predict that significant numbers of people with chronic low-dose arsenic exposure would develop cancer. Understanding the true risk from low-level arsenic exposure is an area of active epidemiological research.

Other heavy metals

Contamination of water with other heavy metals has caused problems in isolated instances. In 1977, the National Academy of Science ranked the relative contributions of these metals in water supplies as a function of man's activities as follows.

> **Very great**: cadmium, chromium, copper, mercury, lead, zinc
> **High**: silver, barium
> **Moderate**: beryllium, cobalt, manganese, nickel, vanadium
> Low: magnesium.

Of these metals, mercury and cadmium are probably the most toxic at the levels found in water. High levels of environmental exposure to mercury occur primarily through the consumption of food tainted by organic (and sometimes inorganic) mercury (see next chapter). However, the uses of mercury compounds that give rise to these exposures, such as the treatment of seeds with phenyl mercury acetate (used for its antifungal properties), can also lead, through runoff, to the contamination of surface water and groundwater. Similarly, short-chain alkyl mercury compounds are lipid-soluble and volatile; therefore they pose a risk of skin absorption and inhalation from bathing in contaminated waters.

Environmental exposure to cadmium has been increasing as a result of mining, refining, smelting, and the use of cadmium in industries such as battery manufacturing. Environmental exposure to cadmium has been responsible for significant episodes of poisoning through incorporation into foodstuffs; however, the same sources of cadmium for these overt episodes of poisoning, such as the use of cadmium-contaminated sewage sludge as fertilizer, can potentially cause contamination of ground and surface water used for drinking and bathing. High cadmium consumption causes nausea, vomiting, abdominal cramping, diarrhea, kidney disease, and increased calcium excretion (which leads to skeletal weakening). As in the case

Chapter 6. Water

of mercury, the toxic effects of chronic exposure to low levels of cadmium are poorly understood. Recent studies have demonstrated an increased rate of mortality from cerebrovascular disease (e.g., stroke) in populations from cadmium-polluted areas.

Pesticides

In today's world, especially in developing countries, the use of pesticides has become inexorably linked with agricultural production. Included under the rubric of "pesticides" are insecticides, herbicides, nematicides (i.e., a type of chemical pesticide used to kill plant-parasitic nematodes), fungicides, and other chemicals used to attract, repel, or control pest growth. Insecticides and nematicides, including the bicyclophosphates, cyclodienes, and the pyrethroids, generally work by inhibiting neurotransmitter function in the peripheral and central nervous systems. Herbicides and fungicides interfere with specific metabolic pathways in plants, such as photosynthesis and hormone function.

Pesticides pose a major threat of contamination to both surface water and ground water. In the United States, approximately 1 billion pounds of pesticides are applied annually to crops. Persistent and broad-spectrum agents such as DDT were once favored. DDT or Dichloro dibenzo trichloroethane is a pesticide. It was widely used after World War 2nd, but it was eventually banned in 1971. Its effect is persistent.   It accumulates in animal fatty tissues and concentrates up the food chain. Aldrin and dieldrin act similarly. DDT was shown to accumulate in the food chain and in living systems, with profound effects, and was prohibited in the United States in 1972; however, it and related chlorinated compounds continue to be used widely outside North America. Moreover, the non-residual and more specifically targeted chemicals and agents that are now in wide use in North America still generate concern because of their long-term effects on ground and surface water.

Highly water-soluble pesticides and herbicides can leach into groundwater; the less soluble, more persistent chemicals can be carried in surface-water runoff to lakes and streams. More than 70 pesticides have been detected in groundwater. Specific chemicals, such as atrazine, are still routinely detected in aquifers and wells.

The most recognized hazard of pesticide exposure is the development of acute toxic effects at high levels of exposure, such as might be sustained by an agricultural worker. The health effects of low-level or prolonged pesticide exposures via drinking water are much less clear. Extrapolation of results from *in vitro* studies to humans suggests the possibility of incrementally increased risk of cancer for many of the pesticides in use. Epidemiological correlations have been found between elevated serum DDT plus DDE, its major metabolite, and subjects who reported hypertension, arteriosclerosis, and diabetes in subsequent years. Of particular concern are recent findings that demonstrate a strong association between breast cancer in women and elevated serum levels of DDE. The overall database of human epidemiological data is sparse, however. In addition, in view of the slower elimination of pesticides in humans and their greater life span, extrapolating toxicity data from experiments on animals to humans may underestimate risks.

The case of aldicarb, a pesticide that has been used widely in recent times in the United States, is illustrative of contemporary issues related to pesticides and groundwater. A carbamate insecticide, aldicarb has been used on a number of crops, including potatoes, which are grown in sandy soil. The combination of the chemical's being applied to soil rather than to plant leaves and the permeability of sandy soil has led to widespread groundwater contamination. Aldicarb has been detected in groundwater in Maine, Massachusetts, New York, Rhode Island, Wisconsin, and other states.

Volatile Organic Compounds

Other very common groundwater contaminants include halogenated solvents and petroleum products, collectively referred to as volatile organic compounds (VOCs). Both groups of chemical compounds are used in large quantities in a variety of industries. Among the most common uses of the halogenated solvents are as ingredients in degreasing compounds, dry-cleaning fluids, and paint thinners.

Chapter 6. Water

Military dumps have recently been recognized for their widespread environmental contamination with solvents.

Historically, once used, these chemicals were discharged directly to land, given shallow burial in drums, pooled in lagoons, or stored in septic tanks. Sometimes the sites were landfills situated over relatively impermeable soils or impoundments lined with impenetrable material; often, however, the sites were in permeable soils, over shallow water tables, or near drinking-water wells. Petroleum products frequently were stored in underground tanks that would erode, or were spilled onto soil surfaces.

These compounds are major contaminants in recognized hazardous-waste sites. For instance, of the 20 chemicals most commonly detected at sites listed on the EPA's National Priority List, 11 were VOCs. Unfortunately, the chemical and physical properties of VOCs allow them to move rapidly into groundwater. Almost all of the above chemicals have been detected in groundwater near their original sites, some reaching maximum concentrations in the hundreds to thousands of parts per million. Once in groundwater, their dispersion is dependent on a number of factors, such as aquifer permeability, local and regional groundwater flow patterns, chemical properties, and withdrawal rates from surrounding groundwater wells.

At high levels of exposure, VOCs can cause headache, impaired cognition, hepatitis, and kidney failure; at the levels of exposure most commonly associated with water contamination, however, cancer and reproductive effects are of paramount concern. Many of these compounds have been found to cause cancer in laboratory animals. In the meantime, regulatory efforts have depended largely on extrapolations from chronic-exposure studies in laboratory animals, in which carcinogens are essentially regulated in terms of concentrations leading to an "acceptable" amount of risk. It is to be hoped that future regulatory efforts will be based on better information (in the form of epidemiological and toxicological studies) and on a better understanding of what constitutes an "acceptable" amount of risk.

Radioactive Contamination

Small quantities of radioactive substances are commonly found in drinking water. Of most concern is radon, which can be found in groundwater in association with geologic formations which are rich in this isotope. Ingestion of radon is probably of little consequence, since it is poorly absorbed and since the alpha radiation it emits is very weakly penetrating. **Aerosolization** is the process or act of converting some physical substance into the form of particles small and light enough to be carried on the air i.e. into an aerosol. The aerosolization of water containing radon, such as occurs in a shower, can lead to inhalation and increased risk of lung cancer. Standards have been set for the amount of radioactivity, including that from radon, which is allowable in drinking water. Water supplies in many communities, including some in the United States, probably exceed this limit.

Of additional concern for some regions of the world are the contamination of surface water and groundwater by radioactive compounds generated by the production of nuclear weapons and by the processing of nuclear fuel. Government secrecy has prevented recognition of many of these incidents for years. For instance, it was only recently acknowledged that radioactive wastes, including plutonium, were dumped directly into the Techa River from 1947 through 1956 at the Russia's weapons-production facility in Chelyabinsk. The Techa was the main source of potable water for thousands of residents of villages along its shores. Similar contamination of surface and ground water has occurred at weapons-production sites in the United States.

Hazardous-Waste Sites and Groundwater Contamination

Many of the specific hazards discussed above threaten water supplies because of their presence at hazardous-waste sites. Epidemiological studies of communities near hazardous-waste sites are plagued by a number of methodological obstacles, some of which were mentioned in the preceding section. Even if studies are performed flawlessly, and an association is discovered, causality is far from proven; moreover, the complex mixtures of chemicals found at most hazardous-waste sites make it exceedingly difficult to pinpoint the culprit substances.

Nevertheless, such studies are vitally important. They provide information on the scope of the problem, and they serve to educate communities about the hazards and the possible (if not exact) risks. Moreover, methods of exposure assessment and outcome ascertainment are constantly improving, as is demonstrated by a study in which slight but significant increases in malformation rates were associated with residential proximity to hazardous-waste sites in New York State.

Pollution and Water Treatment

Remedial action for a contaminated aquifer is complicated, time-consuming, expensive, and often not feasible. If a contamination plume is shallow and in unconsolidated material, excavation and removal is a possible solution; other strategies include *in situ* detoxification, stabilization, immobilization, and barrier formation. Similarly, decontamination of a surface water supply is often complicated by the multiplicity of contaminants involved. Methods of water treatment that might be employed include reverse osmosis, ultrafiltration, use of ion-exchange resins, and filtration through activated charcoal. Clearly, the best solution to the contamination of groundwater or surface water is prevention.

Methods used for disinfecting drinking water can have toxic effects, due to the disinfectants or by their by-products. In the United States chlorine is routinely used, because of its powerful and long-lasting antimicrobial effect and its low cost; however, as a by-product of chlorination, chlorine reacts with substances commonly found in water to generate trihalomethanes (THM), such as chloroform, which increase the risk of cancer. As a volatile organic compound, chloroform can be significantly absorbed through skin contact and inhalation during a shower.

Contamination, water treatment, and expense must be considered in the context of usage patterns. In developed countries, high-quality water is used in huge quantities. In the United States, 50 gallons of high-quality water are consumed per capita per day for domestic uses alone (165 gallons, if one counts commercial uses as well). Less than 1 gallon is actually consumed; the rest is utilized in a myriad of activities, most of which do not require high quality. Approaches to decreasing the use of high-quality water include increased attention to methods of conservation and the institution of dual water systems in which separate plumbing systems deliver high-quality water for culinary use and less pure water for other uses.

Let us look at the **pollution of surface waters**. The number of different industrial and agricultural chemicals that threaten public and private water supplies is enormous. Nitrates, heavy metals, pesticides, and volatile organic compounds are of most concern in terms of human health. The exact nature of the health risks from many of these exposures is not known; this is particularly true with respect to the relationship of low-level chronic exposures to cancer and other long-term effects. Additional epidemiological and toxicological research is important, as are improving risk-assessment methods and defining societal notions of "acceptable" risk. Of equal importance, however, is using existing research to target the prevention of additional contamination of this resource that is so critical to health and survival.

The availability of large quantities of fresh water is very important to our society. Although a human being only needs 2-3 liters of water a day to survive, it requires 1200 gallons of water to produce an 8 oz. steak (but only 165 gallons for 8 oz. of chicken).   EPA classifies waters into Classes I, II, and III, with Class I being the highest quality drinking water and Class III being highly saline water that is contaminated beyond recovery.   Contaminants may be introduced into water from a localized source, such as a single waste effluent pipe or a specific industrial plant. This is known as a "point source." On the other hand small amounts of pollutants may be spread over a very large area and all together may add up to a large influx of pollutant. For example, small amounts of oil may be spread over a large number of parking lots and roads, and during a heavy rainfall

Chapter 6. Water

these will be washed into the storm sewers and eventually find their way into the river. This is known as a "non-point source".

Let us turn to **oil spills**. Crude oil is composed mainly of hydrocarbons with a few percent of sulfur, nitrogen, and oxygen compounds, together with traces of some metals (mostly nickel and vanadium). Crude oil occurs naturally and over geologic time organisms have evolved to metabolize it. However, in an oil spill the crude may be around for several years before it is degraded and there may be considerable short-term damage. Oil spills can occur during transportation and exploration/development, and in addition natural seeps can be locally important. The most serious spills are those that occur at sea because the oil gets widely distributed and containment and cleanup can be very difficult. Other byproducts of oil exploration and production that can be pollutants include salt water and cuttings/mud, among others.

Next let us consider **acid mine drainage**. Sulfide minerals exposed by mining can interact with groundwater to give very acid solutions. Seepages of sulfuric acid solutions with pH in the range of 2.0 to 4.5 have been reported.

The pureness of **lakes** is vital. Because of their relatively small volumes lakes often show a significant response to the various types of pollution.

> 1. **Sewage**. Sewage (together with artificial fertilizers and phosphate) can cause algal blooms and lead to Eutrophication. In this process the decomposing remains of the excessive plant growth deoxygenates the water and it becomes foul-smelling and virtually lifeless.
>
> 2. **Acid rain**. Acid rain is a rain or any other form of precipitation that is unusually acidic, meaning that it has elevated levels of hydrogen ions (low pH). It can have harmful effects on plants, aquatic animals and infrastructure. Acid rain is caused by emissions of sulfur dioxide and nitrogen oxide, which react with the water molecules in the

atmosphere to produce acids. Acid rain has been shown to have adverse impacts on forests, freshwaters and soils, killing insect and aquatic life-forms, causing paint to peel, corrosion of steel structures such as bridges, and weathering of stone buildings and statues as well as having impacts on human health.

3. **Recreation**. Power boats and lakeside homes can be sources of significant amounts of pollution.

Lakes are threatened by a multitude of contaminants. A wide range of synthetic chemicals gets into natural waters from agricultural practices. These include fertilizers, pesticides, herbicides, etc. Industrial processes have also contributed heavy metals, PCBs, and many other long-lasting chemicals that are still in waters and their sediments.

There are other forms of pollution. One is **thermal pollution**. Cities in general are hotter than the surrounding countryside. Local industrial plants may return warmed water from their cooling towers into rivers and streams. Industrial cooling waters that have been warmed up may raise the temperature of streams and rivers. Water released from dams is often cooler than the surface waters in the area.

Another form of pollution is **turbidity** from erosion and other sources. An example would be sediment run-off from clear-cutting. The spreading of salt on winter reads can lead to problems for aquatic organisms that are not tolerant of higher salinities.

It is important to examine the movement of **groundwater pollutants**. Contaminants are moved away from the point of input to the aquifer in a "plume". The exact size, composition, concentration, and shape of the plume depend on the nature of the contaminant, time, the nature of the minerals in the aquifer, the distribution of permeability, and the water flow rate. Major factors controlling the shape and position of the plume are:

1. Water velocity and composition.

2. Density of the contaminating liquid (if it is not miscible).

3. Miscible contaminants tend to move with the water.

4. Adsorption. Different compounds are adsorbed to different extents and so may move at different rates, or may even be removed from solution altogether.

5. Dispersion. Depends on porosity and permeability of the aquifer and the water flow rate.

The configuration of the plume may vary with time, reflecting changes in the rate of contaminant input or water flow rate. In some cases introduction of the contaminating material into the aquifer may be episodic.

Chapter 7. Energy—Oil and Coal

Petroleum

"O for a Muse of fire, that would ascend the brightest heaven of invention."—Shakespeare, (*Henry the Fifth*)

Until about 1800 man-power and animal-power were main sources of energy. Also the burning of wood was a major source, but many forests of the world were being exhausted, and coal began to replacd wood. The use of coal in Western Europe as a source of domestic heating can be traced back through the Middle Ages to the twelfth century. However, it was not until about AD 1500 that coal was widely transported for use as an energy resource; exhaustion of local wood supplies was the prime cause for the change from wood to coal.   Generally, the energy source adopted has been the most convenient in terms of location and use. Before the Industrial Revolution, easily harnessed energy came from the burning of wood, and from wheels and pumps driven by wind or water. The discovery of crude oil in Pennsylvania in 1859 was a major turning point. Today crude oil and natural gas are the leading energy sources.   In order to reduce the extent and momentum of human influences on global change it is necessary to develop and utilize new technologies. In the energy sector, alternative technologies that are renewable and more sustainable are being developed. Economic and social effectiveness of potential policy options must be determined that encourage more sustainable patterns of resource use and management. The burning of wood of energy has far from disappeared. Approximately a fifth of the world's population still relies mainly on wood as domestic fuel. Trees are cut down for cooking and heating purposes.

Coal is abundant and only a few percent of the total has been used. The Earth's coal reserves total many trillion tons; the United States has reserves of at least 400 billion tons.   At the present rates of consumption, coal can last 300 to 500 years. But the best grade of

Chapter 7. Energy—Oil and Coal

the world's coal will be gone much sooner, especially because of the wasteful ways of energy consumption

Petroleum is a complex mixture of hydrocarbons that occur in Earth in liquid, gaseous, or solid form. The term "petroleum" is often restricted to the liquid form, commonly called crude oil, but, as a technical term, petroleum also includes natural gas and the viscous or solid form known as bitumen, which is found in tar sands. In fact, liquid and gaseous hydrocarbons are so intimately associated in nature that it has become customary to shorten the expression "crude oil and natural gas" to "petroleum" when referring to both.

The liquid and gaseous phases of petroleum constitute the most important of the primary fossil fuels. A fossil fuel is a natural fuel formed in the geological past from the remains of living organisms. They are non-renewable natural resources. Biomass is organic material that comes from plants and animals, and it is a renewable source of energy. Biomass contains stored energy from the sun. Plants absorb the sun's energy in a process called photosynthesis. Coal and petroleum also contain stored energy from the sun, but in fossil form.

The nations of the world were "parched" for oil in **1950**. The MIT digital computer (Whirlwind) appears on the scene in **1951. By 1953** geophysicists "with throats unslaked" were "drinking all" of what the digital computer offered, namely digital seismic data processing. However by **1954** cyberanxiety in geophysics took hold. The computer began to be feared. It was deemed that the digital computer was an unreliable omen. In the **1960s**, the old (i.e., pre-computer) methods of finding oil had failed. In other word the old ways could not find oil in except in special favorable circumstances. As a result, most of oil in reservoirs on land and essentially all oil in reservoirs under the sea were unobtainable. Digital seismic processing started its ascendency. Great new oil fields on land and sea were discovered by means of digital seismic processing. All oil shortages were brought on by political forces, and not by an inability to discover new oil fields.

Petroleum became the main source of energy in the world. Every aspect of day-to-day life is somehow influenced by the use of petroleum. Uses of petroleum include: agriculture (running farm machinery, fertilizer, and delivery), transportation, tires, paints industrial power, heating and lighting, lubricants, plastics, pharmaceuticals, petro-chemicals, dyes, detergents. Even sulfuric acid has its origins in the sulfur that is removed from petroleum.

Ironically, now that petroleum could be easily discovered in abundance, petroleum itself began to be feared. However, petroleum represented a force that could not be ignored. By drilling into the crust of the earth, people became knowledgeable of the long and intricate history of the earth. Petroleum (a.k.a. hydrocarbon), which was formed throughout geologic time, is an integral part of the life cycle that makes our planet special. It represents a long history with an abundance of living things, their beauty and wonder. As a battery stores electric energy, world-wide reserves of petroleum store the energy of animal life. This stored animal energy could rejuvenate bacterial life on earth in case of catastrophe, as from volcano activity or asteroid encounter. The digital computer (artificial intelligence) is used to find petroleum (fossil life).

The burning of fossil fuels and biomass releases large quantities of carbon dioxide (CO_2) into the atmosphere. The CO_2 molecules do not allow much of the long-wave solar radiation absorbed by Earth's surface to reradiate from the surface and escape into space. The CO_2 absorbs upward-propagating infrared radiation and reemits a portion of it downward, causing the lower atmosphere to remain warmer than it would otherwise be. This phenomenon has the effect of enhancing Earth's natural greenhouse effect, producing what scientists refer to as anthropogenic (human-generated) global warming. There is substantial evidence that higher concentrations of CO_2 and other greenhouse gases have contributed greatly to the increase of Earth's near-surface mean temperature since 1950.

Modern technology is completely dependent upon petroleum (i.e., crude oil and natural gas). As more and more people worldwide turn to automobiles, more supplies of petroleum will be needed for fuel.

Chapter 7. Energy—Oil and Coal

Some proposals have been made to conserve fossil fuel. Conservation measures and the disincentive effect of highly priced and taxed fuel (particularly in Europe) have reduced the growth of oil imports. Alternative sources of energy were evaluated but found lacking. About 25 percent of total energy consumption (half of all petroleum) goes for transportation. Mass transit can help, but it will mainly affect intra-city transportation and would reduce car use and gasoline consumption by only by a small amount. The modern way of life encourages increased energy consumption. Other sources are nuclear, geothermal, and solar energy, oil-shale processing, and gasification of coal. Oil shale appears to take more energy to produce than is obtained. A major problem is the waste rock from the process whose volume is about twice its original volume. Furthermore, processing the oil shales requires great quantities of water, not available in the semiarid western US where the oil shale is found. A major problem for the coal industry is to find ways to utilize the abundant high-sulfur coals without further damaging air quality. The mining of the widespread coal that occurs in thin layers requires strip-mining. This form of mining alters the landscape and the burning of coal causes pollution.

Resources extracted from the earth, in commercial and industrial terms, are the most fundamental of all natural raw materials used by modern society. Almost all other resources depend on energy for their extraction and processing; moreover, the availability and cost of energy often exerts the predominant control on the economics of metals, bulk materials and, to a lesser extent, water. If, in the future, a cheap and plentiful supply of energy could be made available, then currently uneconomic grades of ore, depths of mining, bulk materials processing and water purification technologies might all become more economically feasible.

Energy resources can be divided into **renewable** and **non-renewable** categories. Renewable energy resources are available in a natural way (such as solar energy) which may be harnessed for human exploitation. Non-renewable energy resources originate from primary energy that has been converted into a stored form (coal, for example) suitable for extraction and reconversion into usable energy. Renewable energy resources, such as the heat of the Sun, are available for all time, and non-renewable energy resources, such as coal, are exploited so much more rapidly than they are formed that their formation is of no concern.

Renewable sources of energy can come from external energy sources such as solar radiation and the gravitational influence of the Sun and the Moon, and from internal energy sources such as the Earth's internal heat and the Earth's rotation and gravity field. Solar radiation is responsible for the atmospheric pressure differences that cause winds (and hence ocean waves), evaporation and re-precipitation of water vapor in the hydrological cycle, and photosynthetic activities among organisms. The thermal energy of the Sun is produced by nuclear fusion. A tiny fraction of the solar power is intercepted by the Earth and absorbed in the atmosphere or on the Earth's surface. The heating effect is greater at the equator than at the poles. This differential heating combined with the Earth's rotation produces parallel belts of high and low atmospheric pressure between which winds blow. Some of the wind energy is converted into water waves through frictional effects at the sea surface. Another small portion of the absorbed solar power appears as gravitational energy in the hydrological cycle, when water evaporated from seas, lakes and so on, returns to the surface as precipitation. Some of this gravitational energy may be harnessed to provide hydroelectric power.

On reaching the Earth's surface, solar radiation encounters surface water, vegetation cover, rocks or soil. A small proportion of this energy is potentially available for consumption as an energy resource either by direct conversion of solar energy or via biological production and conversion. Approximately one-half the energy is used to evaporate surface water. It is thereby transferred back to the atmosphere while the other half is used to heat the surface. Of course, most of this surface energy also is ultimately re-radiated into the atmosphere. However, the atmosphere is a complex system which, because it contains carbon dioxide and water vapor, is more

Chapter 7. Energy—Oil and Coal

'transparent' to incoming short-wavelength solar radiation than it is to outgoing long-wavelength terrestrial radiation. The only way that equilibrium can be established where-by the Earth's heat gain and losses from the upper atmosphere are balanced is for the temperature of the atmosphere to remain at a high mean value of 15°C. If on the other hand the atmosphere were transparent to terrestrial radiation then the equilibrium temperature would be an inhospitable -18°C. The raising of global temperatures due to minor atmospheric constituents is known as the *greenhouse effect.*

Tidal power is exploitable as an energy resource. Some of the energy of ocean tides can be exploited along coastlines where the natural tidal range *(ca.* 1 meter) is accentuated by geographical factors (e.g. tidal amplitudes of 12 meters are recorded in the Bay of Fundy).

The Earth's internal heat occurs both in volcanoes and hot springs. The Earth's interior is hot: it can produce molten rock at temperatures of up to 1250° C and superheated steam. However, these phenomena are mainly confined to several narrow and elongate zones along active continental margins and ocean-ridge zones, the currently active boundaries of the tectonic plates. It is possible to exploit this geothermal heat in areas of recently active volcanoes and other places.

The source of the Earth's internal heat comes from the decay of long-lived radioactive isotopes, the most important being potassium-40, thorium-232, uranium-235 and uranium-238. Of these uranium-235, which comprises only 0.7 per cent of natural uranium, forms the basis of another major category of energy resources: nuclear fuels.

Non-renewable energy resources
The fossil fuels, coal, oil and natural gas, are non-renewable energy resources because they are all forms of stored solar energy. Uranium used as nuclear fuel is also a secondary resource for it represents primary nuclear energy stored when the Earth first formed.   Let us examine the link between photosynthesis and fossil fuels.   All living things on the Earth's surface depend on the trapping of solar radiation by plants. This is true even of higher

vertebrates, which eat either plants or animals that in turn have eaten plants. The processes by which land plants and phytoplankton in the oceans convert carbon dioxide and water into carbohydrate and oxygen is known as photosynthesis. The resulting carbohydrate stores energy until it is broken down by the reverse exothermic reaction, when it gives off heat during combustion, or decay caused by metabolic activity or by some other means.

About 0.2 per cent of the Earth's incident solar power is converted into plants, which together with animals that feed on them are known as the biomass. Photosynthesis continuously creates a biomass with energy content many times greater than the total energy demand, but there are many difficulties associated with converting enough of the fuel to supply our energy needs. Ideally all organic matter is ultimately consumed by decay or combustion so that the intake and output of energy are exactly balanced. The heat balance of the biosphere and of the Earth's surface itself would be disturbed if there were departures from this equilibrium. Over the thousands of millions of years since life appeared on Earth, there have in fact been minor departures from this equilibrium. Some organic matter has accumulated in oxygen-free (anaerobic or anoxic) conditions, such as swamps, bogs and parts of the ocean floor, and this material has not decayed. The existence of fossil fuels, which are the products of non-decayed organic matter, means that there is non-decayed organic matter trapped in sedimentary rocks. Over geologic time, these remains were converted into fossil fuels. The fossil fuel bank that has accumulated over the past few hundred million years represents the most important energy resource. Despite their apparently widespread occurrence, commercial coal fields and oil fields are limited in number.

Renewable energy resources represent a continuously available energy whereas non-renewable energy resources are in limited amounts. The Sun is the most important renewable source of energy. Solar radiation reaching the Earth contributes to winds, waves, atmospheric water circulation, atmospheric heating, and surface water evaporation. The external gravitational influence of the Sun

Chapter 7. Energy—Oil and Coal 159

and the Moon combined with the Earth's axial rotation produces tidal effects in the oceans, which are potentially exploitable sources of power. Various minor constituents in the atmosphere render it less transparent to outgoing terrestrial radiation than to incoming solar radiation. This results in a global average of 15°C for the equilibrium temperature of the surface. A small contribution to the surface energy balance is made by radioactive decay from long-lived isotopes of potassium, thorium and uranium. At depths within the Earth beyond the influence of solar radiation (i.e. greater than a few meters), geothermal energy is most important. Non-renewable energy resources include uranium reserves and fossil fuels. Fossil fuels originated from solar power that has reached the Earth during the past millions of years and was converted into hydrocarbon form. The fossil fuel bank represents that small fraction of the hydrocarbons produced in sufficiently concentrated forms to be commercially extractable.

Coal has two important advantages over wood as an energy resource.   Coal is a more concentrated form of energy than wood and so, for an equivalent energy output, it was cheaper to transport. It also produces a hotter flame than wood, which make the working of iron by blacksmiths easier. One of the most important developments to affect the world energy came with the invention automobile which has created an enormous demand for petroleum products. Most nations have shown a rapid increase in their consumption of oil and gas. The period from 1950 to 1970 saw immense changes in the pattern of world energy supply and demand. Most developed nations changed from being just major coal consumers to being major oil, gas and coal consumers.   Nearly all of them rely heavily on imports of oil and gas from foreign sources, particularly from the Middle East.

Historically, there has been an exponential increase in world demand for energy. Can the fossil fuel bank sustain this demand? There is ample fossil to maintain the production levels well into the next century. But the vast bulk of the fossil fuel bank is locked up in coal reserves, the production of which is difficult to accelerate, whereas oil and gas are in more demand. Can we actually locate and extract the necessary reserves in the vast amounts and at the rates suggested by the projections. To what extent can we reduce our dependence on fossil fuels? Can other sources of power expand to fill the "energy gap" between total demand and fossil fuel availability?

Resources of fossil fuels are unevenly distributed around the world and the reasons are primarily geological. Certain geological conditions were necessary for these resources to have formed. The geological processes that produce energy capital resources depend on critical combinations of environmental and physical conditions coupled with time constraints; for example, coal accumulations only began with the evolution of terrestrial plants. As a result, energy resources are markedly uneven in global distribution and this had been reflected particularly in the sensitivity of oil supply from Middle Eastern sources and the demand for oil in the western world

Historically, wood, wind and moving water have been the most widely used energy resources. But since the Middle Ages, first coal, then oil and natural gas, have become progressively much more important sources of energy. Over 80 per cent of global energy production is based on the use of fossil fuels. With the overwhelming fear of nuclear weapons, it is doubtful that nuclear power will become a growth industry. Following the growth patterns of coal and then oil (plus natural gas) production, much attention is being devoted to renewable energy sources.

The outstanding geochemical characteristic of coal is the remarkably high content of carbon and, to a lesser extent, hydrogen; this is a reflection of the biogenic origin of coal. Coals rich in volatile matter (more than 30 per cent) are easy to ignite and burn freely but with a smoky flame. Low volatile coals are more difficult to ignite but they burn with a smoke-free flame, providing a natural smokeless fuel. Coking properties: the carbonized residue remaining after the volatile matter has been driven off in the absence of air is called *coke*. Coals in the range of 85-89% carbon content (high rank bituminous) become partly fluid on heating and swell up to form a porous coke.

Chapter 7. Energy—Oil and Coal

These *coking coals* are especially valuable for the coke used by the iron and steel industries. Being free of volatiles coke also provides a useful (artificial) smokeless fuel.

The inorganic impurities within coal seams contribute to the residue or *ash* left after combustion. This mineral matter either accumulated with the original plant material or was precipitated during coalification. Its distribution within the seam is important for practical reasons, determining the extent to which the coal can be cleaned before sale. The melting temperature (fusibility) of the ash and its tendency to corrode boiler surfaces are also dependent on the composition of the mineral impurities. These are important considerations when assessing the suitability of coals as boiler fuels or for coke making.

The *clastlc sediments* deposited with the original plant debris are fine muds washed into the accumulating peat deposits and consist of clay minerals and fine quartz particles. Next to the clays, the most important group of impurities is the carbonates. Sulphur is another common impurity found in all coals, in proportions varying from 0.5 to several per cent, and is principally present as pyrite (FeS). On combustion it yields sulphur dioxide (SO_2), which causes corrosion in boilers and contributes to atmospheric pollution. Another important impurity is sodium chloride (NaCl). When it is present in high concentration (over 1 per cent chlorine content) the coal is virtually unusable in power stations because of severe boiler corrosion. Coal also contains minute amounts of a large number of *trace elements,* as do other sedimentary rocks; a few elements may show enrichment in coal, including germanium, arsenic and uranium. Although most of these end up in the ash after combustion, it has been alleged that some trace elements released by burning coal may contribute to atmospheric pollution.

Oil Exploration: Past and Future

After World War II, the US research structure was a self-reinforcing triangle of industry, academia, and government. Today one side of that triangle—industry—is almost absent. The remaining structure is much less stable. Much of the fundamental research in the last century was done in corporate laboratories—Bell Labs, GE Research, IBM Research, and others. Today, only vestiges of these laboratories exist, and they have a much shorter time horizon and are heavily focused on product development. Some would say that the demise of corporate research is the result of the short time horizon of the stock market and/or the demise of the regulated market in telecommunications. It also represents a failure to consider research as an investment rather than an expense—in effect, saying research per se has no lasting value. As a result, instead of developing public policies to encourage corporate research, we are doing just the opposite. It seems that Wall Street financiers, who believe that research has no lasting value, exert a decided influence on research.

Funding for research in physical sciences and engineering has declined or remained flat for several decades. Federal funding agencies have become increasingly risk adverse and have focused increasingly on short term results. The most obvious example is the Defense Research Projects Agency (DARPA), which used to be a shining example of investment in long-term visionary research.

Exploration geophysics is caught up in this maelstrom. However, the danger is even greater. In exploration research, there was never any self-reinforcing triangle. The government has never funded exploration research except for the occasional token, and universities except for a few have generally been unreceptive to exploration geophysics. Historically the oil and geophysical companies alone have supported exploration research. This support was done on a large scale within many dedicated company research laboratories. The oil-company research laboratories of fifty years ago prided themselves to be equivalent to the best of research universities, even as the Bell Labs did. Today those oil company laboratories are gone or greatly reduced in scope, and except for some notable exceptions, which include Exxon and Shell, the research that remains today is done by geophysical companies and by consortia at a few selected universities funded by oil companies. This research is mainly focused on short term results. Without long-term research and the talent that such research attracts, the oil

Chapter 7. Energy—Oil and Coal

industry as well as industry in general will suffer in the difficult times that lie ahead. The geophysical companies, large or small, have always been at the forefront of research. Government agencies would do well to give geophysical companies the support necessary to do the long-range research that must be done to keep the wheels of industry turning.

Exploration geophysics, as well as the oil industry, are swept up in a torrent of events that is engulfing the world today. General speaking, every industry goes through five major phases.

> The **first phase** (Act 1), the setting, is dominated by the pioneers who prepare the way for others.
>
> The **second phase** (Act 2), the rising action, is dominated by the inventors who devise new contrivances and ways of doing things.
>
> The **third phase** (Act 3), the climax or the peak, is the turning point. It is dominated by the managers who nurture efficient production and distribution.
>
> The **fourth phase** (Act 4), the falling action, is one of decline; it is dominated by financiers who consolidate, downsize, and merge companies.
>
> The **fifth phase** (Act 5), the denouement or resolution, is dominated by lawyers and politicians who rationalize the actions and straighten out misunderstandings. The words of a popular song are: **"We are driving in the sun, working hard and having fun. California, here we come, right back where we started from."**

Let us now look at exploration geophysics in particular.

> In **Act 1**, the pioneers were the early geologists and geophysicists who showed how the study of sedimentary rock layers would indicate the sources of petroleum.

In Act 2, the inventors were the hard working geophysicists who devised better instrumentation and turned to computers to do the digital processing required for the production of beautiful seismic images that unlock the secrets of the underground rock layers.

In **Act 3**, the managers were business men who turned exploration into a fine-tuned and efficient operation by the intelligent use of three and four dimensional seismology. Is exploration geophysics entering Act 4? A national oil company (NOC) is a state-owned integrated oil company that controls all or most of the petroleum industry in that nation to the exclusion of other companies.

In **Act 4**, the financiers are the executives who say that finding oil in the ground is passé; instead oil is bought from a national oil company, or by merging, or by acquiring an oil company on Wall Street, or by purchasing oil from traders on the ABC islands, or elsewhere. The financiers dismiss the research departments and replace experienced geophysicists by lesser paid personnel. They downsize companies that took years to build up. They indulge in mergers and acquisitions which can be described as a type of cannibalism. When an industry collapses, the more powerful companies seize the assets of the weaker companies.

Act 5 finally comes. The politicians say that petroleum is the problem and not the answer. They tell us that fossil fuels are the main cause of global warming and all the bad things that it entails. Pogo, the beloved cartoon creation of Walt Kelly said, "We have met the enemy and he is us." The explorationists discovered Pandora's Box, and opened it. If the geophysical discoveries of petroleum were never made, then aspiration levels would never have reached their present heights, and the world would not be in an energy crisis today.

Chapter 7. Energy—Oil and Coal

Most people would describe the energy crisis as follows. Fossils fuels (i.e., coal and petroleum, which includes both crude oil and natural gas) are used to produce the great bulk of the world's energy. The world will greatly increase its population and its use per capita of energy in the years ahead. The combustion of fossil fuels produces carbon dioxide, which acts as a greenhouse gas in producing global warming. As a result the development of carbonless forms of energy is desired. The four realistic options of reducing carbon dioxide emissions from electricity generation are:

1. Increase efficiency

2. Expand use of renewable energy sources

3. Capture carbon dioxide emissions and sequester the carbon

4. Increase use of nuclear power

Nuclear power is costly and dangerous. It is faced with four critical problems, namely (1) cost, (2) safety (3) waste, and (4) proliferation. Even if the first three were overcome, the current international safeguard regime is inadequate to meet the security challenges presented by global nuclear expansion. The reprocessing system used in Europe, Japan and Russia involves separation and recycling of plutonium. Such a system presents such an unwarranted risk for the proliferation of nuclear weapons that no new nuclear energy plants should ever be built anywhere. To emphasize the seriousness of this problem all existing nuclear plants should be phased out. Turn down the lights. Better some pain now than nuclear destruction.

Renewable resources such as hydro, wind, solar, biomass, and geothermal are more amenable to fixed uses such as heating and lighting. Transportation, on the other hand, requires a fuel that can be carried on the moving vehicle. Unfortunately Thomas Edison failed in his quest for a storage battery that could adequately power electric vehicles. As a result hydrogen, not stored electricity, has become the carbonless fuel of choice for transportation. A tremendous amount of hydrogen would be needed to satisfy transportation needs. Nuclear can produce electricity, which in turn can produce hydrogen, but nuclear must be ruled out. Coal is a massive energy resource that has the potential for producing cost-competitive hydrogen. However, coal processing generates large amounts of carbon dioxide. In order to reduce these emissions, massive amounts of carbon dioxide would have to be captured and safely and reliably sequestered for hundreds of years. The commercialization of a large-scale, coal-based hydrogen production option (and also for natural-gas-based options) requires much research and development. "**Dirty old coal, here we come, right back where we started from.**"

For transportation, current opinion favors the use of fuel cells. A battery is a connected group of electrochemical cells that store electric charges and generate direct current. A fuel cell takes the place of a battery. A fuel cell is an electrochemical generator that produces direct current from a chemical reaction. In the usual case, the chemical reaction is one that combines oxygen and hydrogen. The oxygen comes from the air. The hydrogen is stored on the vehicle, and acts as the fuel for the fuel cell. Hydrogen is the lightest and the most abundant chemical element in the universe. However, it is found only in trace quantities in the observable portion of our atmosphere, only about 0.00005 percent by volume of dry air.

The hydrogen used in fuel cells is an energy carrier, and must be manufactured using energy from other sources. Let us review the steps taken to power a motor vehicle. A source, for example hydro power, is required to make electricity. The electricity is then used to produce hydrogen. The hydrogen is then stored on the vehicle. The fuel cell uses the hydrogen to generate electricity. The electricity drives an electric engine which produces the mechanical energy to move the vehicle. In other words, the kinetic energy of moving water is transformed first to electricity, then to hydrogen, then to electricity, and finally back to kinetic energy. One must be cautious of the explosive nature of hydrogen. In Massachusetts on May 6, 1937 my mother, brother, sister and I looked into the sky and watched the

Chapter 7. Energy—Oil and Coal

German airship Hindenburg float across the blue. It was the most beautiful sight I had ever seen. Not many families could afford radios then, so news was usually delayed. I remember well that my mother had a telephone call from her sister Madeline that very evening. Madeline said that she had heard that the Hindenburg had exploded. My mother responded: No, I saw the Hindenburg a few hours ago and it is fine. Today people are, in effect, repeating my mother's statement. The Hindenburg is fine; hydrogen is safe. "**Pretty airship, here we come, right back where we started from.**"

There are major obstacles that prevent the use of hydrogen as a fuel. The transition to a hydrogen economy requires many technological innovations. There must be dramatic progress in the development of fuel cells, storage devices, and distribution systems. Fuel cell lifetimes are much too short and fuel cell costs are much too high. A major problem is the high cost of distributing hydrogen to dispersed locations. Problems concerning security, environmental impact, and safety of hydrogen pipelines and dispensing systems are close to insurmountable. Because hydrogen is the lightest of elements, it takes a lot of effort to keep hydrogen from escaping from a container. A complex set of seals, gaskets and valves is needed, and still liquid hydrogen tanks boil off about three percent of the stored hydrogen a day. All of these hurdles must be overcome before there can be widespread use. For example, platinum is a large contributor to the cost of fuel cells, especially the proton exchange membrane (PEM) fuel-cell which is being considered for diverse applications, including automobiles. The short lifetime of a fuel cell requires a constant renewal of the platinum. In ways, one may think that the platinum, not consumed but lost, is the true fuel used by the fuel cell, and that the hydrogen is just an insignificant adjunct. Platinum is a precious metal. Any massive demand for platinum would increase its cost to astronomical levels. Just as the invention of the silicon chip made it possible for the average person to own a computer, several comparable breakthroughs are required before the average person can look forward to a fuel-cell automobile. Governments and industries are expending billions of dollars in this effort. Success is not certain.

The key word is the acronym **EROEI**, which stands for "**Energy Return On Energy Invested**." More energy is used to produce, store and transport hydrogen than can ever be gotten back when the hydrogen is used a fuel. For this reason, no one ever computes the EROEI for hydrogen. This is the price one pays for the luxury of transportation. Ethanol is another alternative fuel that is an energy sink, so again the EROEI is never mentioned. In fact, the so called alternatives to oil are, by and large, derivatives of oil. Massive amounts of oil and other scarce resources are required to locate and mine the raw materials (steel, silver, platinum, copper, and uranium) necessary to build, maintain, and dispose of fuel cells, solar panels, windmills, and nuclear power plants. Without an abundant and reliable supply of petroleum there is no way to scale these alternatives to the degree necessary to replace petroleum as fuel. For the foreseeable future the world will remain dependent for its energy supplies on fossil fuels (unless there is a massive proliferation of nuclear plants, which is unacceptable).

Natural gas and condensates are, of course, forms of petroleum. So are the Canadian oil sands. However, oil derived from these oil sands is financially and energetically intensive to extract. Conventional oil enjoys an EROEI of about 30 to 1. The EROEI for oil sands can be as low as about 1.5 to 1. In other words, it would cost about twenty times as much to produce the same amount of oil from oil sands as from conventional reservoirs. The problems for oil shale are even worse than for oil sands. From the large scale use of oil shale, the environmental damage to the land and water may be insurmountable in the long run. Coal will still be around when all the economic oil is used up. And finally there is wood and hay. My father was born in Missouri in 1872. Although the first oil well had been drilled in Pennsylvania about ten years earlier, my father only knew about renewable fuels: wood for the fire and hay for the horses. "**Our dear old farm, here we come, right back where we started from.**"

The law of diminishing returns says that an increase of input to a either a fixed or exhaustible resource will cause the output to increase for a time, but after a point the extra output will become

Chapter 7. Energy—Oil and Coal

less and less. The outlook for future petroleum reserves has always been disheartening. But the more important point is that the law of diminishing returns does not just apply to oil. More generally, we may think of the earth as the fixed resource and people as the variable resource. As we apply more and more people to harvesting the goods of the earth, the output at first increases, but in time the extra output becomes less and less. In fact, it may turn out that oil may not be the most basic need; instead it may be arable land and water.

Shakespeare's King Lear says to his two eldest daughters, "O, reason not the need." In other words, Lear says: Don't try to apply rational calculations to need. Not only is the world's population going up exponentially, but so is each person's need for energy. O, reason not the need. From the study of paleontology indicates that many species have experienced exponential growth for a certain time span, but everyone has either crashed or leveled off..

In 1953 the outlook for oil exploration was bleak, as usual. Of course, everyone trained in geology knows that oil will eventually be exhausted. The question is when. In 1953 there was a general belief that exploration has attained extremely high levels of importance in the mining field, especially for petroleum, where current rates of oil and gas production have soared to all time maximums. Expenditures for exploration, wildcat drilling, geological and geophysical work over the 11 year period from 1937 through 1947 increased by more than four times. Nevertheless, only a near-constant value for the ratio of annual production vs. proved reserves was maintained during this period."

In 1953 a good part of the earth's sedimentary basins, including essentially all water-covered regions, were classified as no-record (NR) areas. The reason was that the raw exploration seismograms from these areas yielded no visible reflections and so could not be interpreted. Yet the decades of the 1940s and 1950s were replete with inventions, not the least of which was the modern high-speed electronic stored-program digital computer. Almost every major oil company except Shell joined the MIT Geophysical Analysis Group in

1953 to use the digital computer to process the NR seismograms. The task of the computer was to remove interference such as ghosts, reverberations and other multiple reflections so as to yield the underlying primary reflections. In any case, the Shell Development Company in 1953, perhaps out of curiosity, invited me to their Bel Air Laboratories in Houston. Their geophysicists were the best. Shell had everything: money, equipment, resources, and people. On my side I had nothing but enthusiasm in the belief that oil could be found in NR areas though the use of digital processing.

At the end of my visit I was shown into the office of M. King Hubbert, the Chief Consultant (General Geology) of the Shell Development Company. There we two were alone. Immediately he started throwing statistics at me: decline rates, expenditure rates, success rates, and wanted me to answer the question of when the peak oil production would be reached. I was not used to verbal mathematics. I needed written equations and graphs, but no paper or blackboard was available. (Sven Treitel told me that at one time, the chief patent attorney at Amoco thought that everything had to be committed to paper for patent purposes, so all the blackboards were removed from the Amoco Research Center. Luckily Sven was able to talk him out of that seemingly brilliant idea.)

Hubbert had position, experience, knowledge, authority; I had nothing but a blind faith in the use of computers in oil exploration. It was like a courtroom. Hubble sat in the judge's chair, an imposing greater-than-life figure behind his large desk. Alone I sat on the cold witness chair. Hubbert's reasoning was based on a "ceteris paribus" (all else remaining the same) assumption. But all else was not the same. I believed that digital computers would change the scope of exploration. However, nearly everyone in authority in universities and in industry said that computers would never work. I argued that a Gaussian curve is symmetric with a central peak. But the oil recovery curve most likely is skewed. The leading edge of the curve is in the past, so it cannot be changed. But the peak and the tailing edge are in the future and they can be changed. If geophysics accepts the cries of the pessimist, then the peak will come soon and the

Chapter 7. Energy—Oil and Coal

tailing edge of oil production will fall off sharply. The resulting production curve will have a negative skew. On the other hand, if geophysics goes digital, the peak will be driven forward and the tailing edge will fall off slowly. The resulting production curve will have a positive skew. Total oil production will be much greater. If only I had a blackboard I could have drawn the two curves. The interview quickly turned to other things and soon was over. When I walked out of Hubbert's office, I had that oppressive feeling that everything I said fell on deaf ears. I was used to that. Few people took computers seriously. Fortunately the great Shell geophysicist Aaron Seriff was there to console me and he made me feel better. (Eleven years later, Shell Development Company came to me in Cambridge, Massachusetts, and gave Geoscience Inc., the company in which I was vice president, a large contract to accelerate their entry into the digital world.)

Hubbert's predictions, revised so that the peak is sometime between 2005 and 2008, have come true, at least that is the consensus of opinion. However, the forty years of rising oil production from 1965 to 2005 would not have come true if there had not been digital. The reason is that digital seismic processing made possible the exploration of no-record (NR) areas, all of which in 1953 were off limits to exploration. For example, oil production from the NR area known has the Gulf of Mexico is a major contribution to America's output. The oil production of Britain and Norway, and the natural gas production of the Netherlands, come from the NR area known as the North Sea. In fact most of the oil discovered in the past forty years comes from NR areas, all made possible by digital.

Many people think that we have now arrived at the Hubbert peak, or are very close to it. After we reach the peak, oil production will inexorably decline. This situation is exacerbated by the fact that the use of petroleum is rising exponentially because more and more nations want to join America as a profligate consumer of oil. As geophysicists we know that we can find more oil, much more than the financiers believe can be found. I believe that we can extend the Hubbert peak to years well beyond the year 2008, if exploration geophysics is given the chance. For example, the North Sea has already outlived the original predictions of its decline by more than a decade and will continue to provide valuable volumes of both oil and gas for many years to come. At the same time, we must practice conservation. We must husband oil and all of the earth's precious resources. We must continue with the development of alternatives.

We must take the advice of Dr. William A. Wulf, namely, instead of developing public policies to discourage corporate research, we must do just the opposite. Government should ease the way for the oil and geophysics industry to play a leading role in the solution of energy problems. Geophysics will continue to explore for the sources of the metals and minerals needed for alternative energy. Geophysics stared the big bad wolf (the Hubbert peak) in the face in 1953, and then began to use computers on a large scale and found great new reserves of petroleum. Geophysics started out as the largest user of computers, and held this distinction for decades until finally government exceeded geophysics in usage. The digital images of our planet, inside and out, come from the digital processing methods originated in geophysics. These same methods are now being used by nearly all other scientific and engineering disciplines. They are especially used in medicine to produce the digital images of the interior of the human body which are revolutionizing health care. Geophysics has provided digital methods for the non-destructive testing required to insure that airplanes and machines are safe to operate. Geophysics is needed to clean up the environment and to properly dispose of nuclear wastes. Geophysics is needed to assure clean water supplies and to make the dessert bloom. And geophysics will again stare the big bad wolf in the face and find new supplies of oil, and also find ways to make oil left in the ground economically recoverable. Geophysics is the science best qualified to find methods to ameliorate global warming.

Dr. Sven Treitel says, "The remaining oil and gas in the ground will be much harder to find and produce in comparison to what has already been found and produced.   Yet the oil industry has chosen precisely this point in time to gut its R&D capabilities in deference to

Chapter 7. Energy—Oil and Coal

the Wall Street analysts, who are clueless when it comes to technical issues. R&D at the oil companies basically came to a grinding halt during the early nineties. Because the lead-times for the implementation of radically new methods are of the order of at least a decade, much technology that could have been developed in time to cope with the increasingly formidable technical challenges will never see the light of day." It is time to give exploration geophysicists the resources and the authority necessary to proceed on its appointed tasks. Exploration geophysics performs the vital work that no one else can do, and the dedicated people who make up exploration are the most valuable asset of all, bar none. And so we face the future on this note: "**Geophysics here we come, right back where we started from.**"

Chapter 8. Nuclear Energy

"And though she be but little, she is fierce."—Shakespeare (*A Midsummer Night's Dream*)

Units for Measuring Radiation

Even a little radiation is fierce. Standard units have evolved that are used to measure radiation and its impact on material.

The **Roentgen** is an obsolete unit of radiation dosage that describes the electromagnetic field associated with radioactive decay. The Roentgen is equal to the quantity of ionizing radiation that will produce 1 electrostatic unit of electricity in one cubic centimeter of dry air at standard temperature and standard atmospheric pressure. It corresponds to a quantity of gamma rays that deposit 87.7 ergs per gram of air at standard temperature and pressure.

A **curie** (Ci) is a measure of total radioactivity or source strength and is equal to 37 billion disintegrations per second. A Curie represents the radioactivity of one gram of radium. The source strength in curies is not sufficient for a complete characterization of a source; the nature of the element (for example, Pu139, U138, Sr90) and the type of emission (for example, alpha) are also necessary. The **Becquerel** (Bq) is also used as a measure of source strength, 1 Bq = 1 disintegration per second; one curie =37 billion Bq.

A **rad** (for radiation absorbed dose) is that quantity of ionizing radiation that leads to absorption of 100 ergs/gram of absorbing material. A complex relationship exists between the quantity of radiation, the curie, and the radiation dose rate, rad/sec. This relationship depends on the energy of the radiation, the type of radiation, the path of the radiation, and the amount of absorbing material between the emitter and the receptor.

The **quality factor**, called relative biological effectiveness, or RBE, describes the biological effect being considered. All types of ionizing radiation do not produce the same biological effects for a given amount of energy delivered to human tissue. Radiation

Chapter 8. Nuclear Energy

yielding higher specific ionization along its track will generally produce a greater effect, but the quantitative difference will depend on the tissue or organ and biological change in question. Different types of ionizing radiation, even for a given amount of energy, produce different effects on living tissue. The quality factor takes into account the differing biological effects of alpha, beta, and gamma radiation. Quality factors are determined for chronic, low-level doses of radiation, using effects that occur in an individual during a lifetime of exposure. Acute, high-level doses produce different effects and have different quality factors.

The **rem** (for Roentgen equivalent man) represents dose equivalent. The rem is the product of the dose in rad and the quality factor RBE. (The quality factor is given as QF = rem/rad.) This unit represents the actual ionizing effect on the human body. A **millirem** is one-thousandth of a rem. One millirem would be the dose from one year of watching color TV for a few hours per day. For each one millirem of radiation a person receives, it is calculated that his risk of dying from cancer is increased by about one chance in ten million. Dose equivalent is also expressed in **sievert** (Sv)., 1 Sv = 100 rem.

Doses are also often expressed in terms of **population dose**, which is measured in person-rem. The population dose is the product of the number of people affected and the average dose in rem. That is, if a population of 100,000 individuals receives an average whole-body dose of 0.5 rem, the population dose is 50,000 person-rem.

The amount of natural, background radiation varies greatly from place to place. In the United States, the average annual natural or background radioactivity is about 300 to 350 millirems, but the range is large; it may be as low as 60 millirems or as high as 600. Extensive research has established that it takes exposures well in excess of 100,000 millirems for a detectable effect. Cancer will result in half of the cases at exposures of 400,000 millirems. The natural background is typically:

      consumer products   10 millirems per year

medical procedures	80 millirems per year
rocks and soils:	20 millirems per year
building materials	10 millirems per year from
internal Potassium-40	25 millirems per year
radon	180 millirems or more
cosmic rays	30 millirems per year

The totals are about 335 millirems per year for each person. Some places will have more natural radiation than other places, depending upon such factors the amount of uranium-thorium concentration in the soil.

The Energy Perspective and Nuclear Power

Consumption of energy has been increasing rapidly due to growing population, rising per capita consumption, rising productivity, use of throwaway products, and waste. In addition to its use as a fuel, oil and natural gas are needed as starting material for the production of fertilizer, plastics, and other important chemicals. Petroleum usage can be reduced by conservation and by substitution of alternative energy resources. At present, fossil fuels (coal, oil, and natural gas) supply approximately 90% of the total U.S. energy consumption, with oil and natural gas responsible for more than 70%. Only for the generation of electricity are there viable alternative energy sources. Every use category except the generation of electricity is presently totally dependent on fossil fuels. For the generation of electricity both coal and nuclear fuels are viable substitutes. The hydropower resource, being dependent on rainfall, is inadequate to substitute further on any significant scale.

Solar energy may be used more extensively for hot water and space heating and cooling within a few years. It is almost economically competitive in various locations; its prospects will improve with rising natural gas costs. But it now seems most unlikely that solar energy will be an economical way to generate electricity. Geothermal

Chapter 8. Nuclear Energy

energy is being used for electricity generation, but its current use is limited to dry steam fields which are quite rare. Thus neither solar or geothermal can grow fast enough to be a significant resource for the next 20 years.

The Chain Reaction

Fission is the **splitting** of the nucleus, whereas **fusion** is the **joining** of two nuclei. Fission requires considerably less energy than fusion to carry out, but also releases less energy than fusion. Both are used in nuclear weaponry. **Nuclear fission** is a nuclear reaction or a radioactive decay process in which the nucleus of an atom splits into 2 smaller, lighter nuclei. The fission process often produces gamma photons, and releases a very large amount of energy even by the energetic standards of radioactive decay. **Nuclear fusion** is a reaction in which two or more atomic nuclei are combined to form one or more different atomic nuclei and subatomic particles (neutrons or protons). The difference in mass between the reactants and products is manifested as either the release or absorption of energy. This difference in mass arises due to the difference in atomic "binding energy" between the atomic nuclei before and after the reaction. Fusion is the process that powers active or "main sequence" stars, or other high magnitude stars.

Nuclear power is a product of nuclear fission or the splitting of atomic nuclei, the principle of the A-bomb. Einstein showed that mass was equivalent to energy, opening up the possibility that matter might serve as a potent source of energy. Fission is a special type of nuclear transformation possible only in very heavy nuclei, such as those of uranium. Fission, made possible by neutrons, releases an immense quantity of nuclear energy. While most nuclei changed only slightly after absorbing a neutron, the uranium nucleus changes by a large amount; it splits in two. The combined mass of the parts into which the uranium nucleus splits is less than the mass of the original uranium nucleus. According to Einstein's equation, this decrease in mass is converted into energy. The energy is present in the form the kinetic energy of the fragments of the uranium nucleus, which separate at tremendous speed.

The neutrons that cause fission in heavy nuclei make possible the chain reaction. The chain reaction is a process that, once started, can continue to release nuclear energy. Let us explain. The uranium nucleus contains very many neutrons. As the uranium nucleus splits, not only is energy released, but several additional neutrons are emitted. The emission of neutrons, along with energy, during the act of fission makes nuclear energy practical. For in fission we start with one neutron and end with the release of a large amount of energy plus several neutrons, usually two or three, in place of the original one. The uranium nucleus is disintegrated in the process.

A reaction based on fission would spread by itself, releasing more and more energy, a method of releasing the enormous stores of energy held in nuclei in the form of mass. Fission of U235 can be thought of as occurring in steps: first, absorption of a neutron in the nucleus; second, the agitated nucleus, with the extra neutron; third, the beginning of fission; finally, fission with the emission of two or three neutrons.

Fission can produce a reaction that sustains itself—in other words, a nuclear chain reaction. The neutrons emitted during the fission of one uranium nucleus cause other uranium nuclei to split, still more neutrons are produced, which tap the energy of still more nuclei. This process leads to enormous numbers of neutrons very quickly. However it is by no means certain that the neutrons produced in the fission of uranium will cause other uranium nuclei to split. The difficulty arises from the fact that a neutron penetrates extremely easily through matter, having small chance indeed of scoring a direct hit on a nucleus. And only if the neutron collides directly with the nucleus of the uranium atom can it cause fission and production of more neutrons. Let us imagine an experiment in which we are shooting neutrons at a piece of uranium that is so small that a chain reaction cannot take place. Now and then a neutron will split a uranium nucleus, but the neutrons thereby produced will almost certainly pass out of the small piece of uranium and nothing further will happen. To attain success the size of the lump must be increased. However, it is impossible to say, without a great deal of additional

Chapter 8. Nuclear Energy

information, how large the lump must be in order to keep neutrons from escaping and thus attain a chain reaction. It might even be possible that the chain reaction would not occur however much the lump of uranium should be enlarged. Experiments show that it was the isotope U235 that split, not the much more abundant U238. As a result, as was soon learned, the chain reaction could never be attained in a lump of normal uranium metal, regardless of size. For the pure U236 isotope, measurements revealed that the chain reaction would be attained if the lump should be larger than a certain "critical size, several inches in diameter. An amount of U235 less than critical will not support the chain reaction and is a harmless, nonexplosive piece of metal. But once the lump is increased beyond the critical size, things are far different. Should a fission occur, the resulting neutrons will not escape, and their number as well as the rate of energy release will increase rapidly. If the U235 should be well over the critical amount, an explosion of tremendous violence would result. Sub- and super-critical masses of uranium behave differently after absorption of a neutron. The process is shown schematically here. Many more collisions occur in the super-critical lump.

The products of the chain reaction are the same whether applied in the atomic bomb or, in a controlled way, in a nuclear reactor. They are (1) energy, which results from the conversion of mass and appears as energy of motion of the fission fragments, (2) fission fragments, usually very unstable and hence intensely radioactive, and (3) neutrons, emitted copiously in all chain reactions. The fission chain reaction proceeds at extraordinary speed. The release of energy from the disappearance of mass in enormous numbers of fissioning nuclei would produce large amounts of heat almost instantaneously. When such large amounts of energy are produced in a limited space and in a short time, we have, of course, an explosion of enormous magnitude. Such a chain reaction results in the atomic bomb. An amount of U235 definitely larger than the critical size will inevitably and immediately explode violently if quickly assembled into a compact mass. There always are some neutrons present in uranium so that a chain reaction will proceed immediately once the assembled mass becomes supercritical.

Let us now examine the controlled chain reaction In a controlled chain reaction it is not necessary to use the single isotopes U235, as in the bomb, but simply the element as it occurs in nature—about 99 per cent U238 and 1 per cent U235. The reaction took place in what is now called a nuclear reactor. For example, a reactor can be made from a pile of graphite blocks interspersed with lumps of uranium. Graphite is chosen for the pile material because carbon atoms of which it is composed do a very efficient job of slowing down neutrons, but have a relatively small tendency to absorb them. Thus the presence of graphite ensures that a large percentage of the neutrons produced in the fission of uranium will themselves go on to produce other fissions. The pile, like the components of an atomic bomb, has a critical size, but its critical size, instead of a few inches, is about fifteen feet. When the pile was made larger than critical, it began to chain react, but did not explode. The reason is that the chain reaction increases in intensity extremely slowly if the amount of material is barely above critical. The structure was built with great care to assure just such a slowly increasing reaction rate. It was found to be very easy to control the rate of the chain reaction, or power of the pile, by simply inserting a cadmium rod into the pile to regulate it. Cadmium is such a strong neutron absorber that, upon insertion, it would stop the chain reaction completely. The pile power could be held at any desired level by simple motions of the cadmium rod, called the "control rod." Although the low-power chain reaction was easy to produce, once pure uranium and graphite were available, a long road yet remains before there is widespread and practical production of electric power from nuclear reactors. The power produced in the first chain-reacting pile appeared as heat; the graphite and uranium of the pile became hotter as the chain reaction continued. But in order to get useful power, say as electricity, from the chain reaction, the great quantities of heat energy must be removed and transformed into electrical energy. Present-day reactors are far different from Fermi's simple pile.

Chapter 8. Nuclear Energy

Having been studied and developed over many years, the process of converting heat energy into steam, then into electrical energy in a turbine, is highly advanced. But in converting the heat of nuclear fission there are many problems in addition to those arising in the production of electricity by the burning of coal. In any chain reaction, as we have seen, one of the products is the radioactive fission fragments. Thus inevitably in any atomic power plant large amounts of radioisotopes, with enormous radiation intensity, will be produced. The problems associated with the safe handling of this radioactivity are new and formidable problems in the generation of electrical power on a broad and inexpensive scale. When heat is produced in a nuclear reactor, the radioactivity is so intense that it is all but impossible to repair equipment. Because of the great hazard of radioactivity it is necessary, although difficult and expensive, to design a plant that will not need repairs for years. Because of the problems inherent in the dangers of radioactivity, atomic power is still more expensive than power produced from the burning of coal or water power.

To date we can use only one of the atomic isotopes, uranium 235, as fuel in our burner reactors; this isotope constitutes less than 1 percent of natural uranium. However it is possible to bombard U-233 and Th-232, to produce the isotopes Pl-239 and Th-233, which can then be used as nuclear fuel.

High-grade uranium deposits are scarce, but many widely distributed rock units such as some black shales and some granitic plutons are essentially low-grade ores and contain large quantities of uranium or thorium. According to data from the United States Atomic Energy Commission (now the Energy Research and Development Administration), world demand for uranium is expected to increase at an average rate of 15 percent per year for the next 20 years. This is a greater increase than any other mineral has ever undergone, and it will require many geologists and great expenditures to keep up with demand.

As we know, however, disposal of the nuclear wastes from the fission process is a serious problem. The actual amount of wastes can be small. When solidified the annual wastes from a 1-million-kilowatt nuclear power plant can be compressed into 1 cubic meter. But, what do we do with it?

Breeder reactor

Natural uranium contains only seven atoms of uranium-235 out of a thousand. The rest are uranium-238. Uranium-235 is a practical nuclear fuel. That is, slow neutrons will cause uranium-235 atoms to undergo fission (break in two) and produce more slow neutrons which will bring about further atomic fission and so on. Uranium-233 and plutonium-239 are practical nuclear fuels for the same reason. Unfortunately, uranium-233 and plutonium-239 exist in nature only in the barest traces, and uranium 235, though it does exist in appreciable quantities, is still rather rare.

Uranium-238, the common variety of uranium, is not a practical nuclear fuel. It can be made to undergo fission but only by fast neutrons. The uranium-238 atoms that break in two produce slow neutrons and these do not suffice to bring about further fissions. Uranium-238 can be compared to damp wood, which may be set afire but will eventually fizzle out. However, suppose uranium-235 is separated from uranium-238 (a rather difficult job) and is used to set up a nuclear reactor. The uranium-235 atoms that form the fuel of the reactor undergo fission and send out vast myriads of slow neutrons in all directions. The reactor can be surrounded by a shell of ordinary uranium (which is mostly uranium-238). In such a case, the neutrons entering that shell will be absorbed by the uranium-238. The neutrons cannot force the uranium-238 to undergo fission, but they will bring about other changes which, in the end, will produce plutonium-239. If this plutonium-239 is separated from the uranium (a rather easy job), it then can be used as a practical nuclear fuel.

A nuclear reactor that breeds new fuel in this manner to replace the fuel that is used up is a "breeder reactor." A breeder reactor of the proper design will produce plutonium-239 in quantities greater than the uranium-235 consumed. In this way, all earth's supply of uranium, and not just the rare uranium-235, becomes a potential

Chapter 8. Nuclear Energy

fuel supply.   Thorium, as it occurs naturally, consists entirely of thorium-232. This, like uranium-238, is not a practical nuclear fuel, since it requires fast neutrons to make it undergo fission. However, if thorium-232 is placed in a shell around a nuclear reactor, the thorium-232 atoms will absorb neutrons and, without undergoing fission, will eventually become atoms of uranium-233. Since uranium-233 is a practical fuel which can be easily separated from thorium, the result is another form of breeder reactor, one which makes the earth's supply of thorium available as a potential nuclear fuel.   The total quantity of uranium and thorium on earth is about 800 times as great as the supply of uranium-235 alone. This means that the proper use of breeder reactors could increase earth's potential energy supply through nuclear fission power plants 800-fold.

Commercial reactors in the United States are either pressurized water reactors in which the water that removes heat from the nuclear reactor core (the "primary coolant") is under pressure and does not boil, or boiling water reactors in which the primary coolant is permitted to boil. 3 In any case, the primary coolant transfers heat to the steam system (the "secondary coolant") by a heat exchanger that assures complete physical isolation of the primary from the secondary coolant. A third cooling system provides water from external sources to condense the spent steam in the steam system.

All thermal electric power generation produces large quantities of waste heat. Fossil fuel electric generating plants are at best about 42 percent thermally efficient: That is, 42 percent of the heat liberated by combustion in the boiler is converted to electricity, and 58 percent is simply dissipated into the environment. By comparison, nuclear plants are 32 percent efficient at best.

Sources of radiation
Radiation is energy traveling in the form of waves, particles, or bundles of energy called photons. Some everyday examples are microwaves used to cook food, radio waves for radio and television, radar for location and tracking of vehicles, X rays used in medicine and dentistry, and sunlight.

Radioactivity is a natural and spontaneous process by which the unstable atoms of an element emit or radiate the excess energy of their nuclei as particles or photons and change (or decay) to atoms of a different element or to a lower energy form of the original element. The second element may also be radioactive and will continue the progression toward lower energies in a cascade of different elements (or decay chain) until a nonradioactive, stable form is reached.

Radioactivity is a property of some materials, while radiation is energy that is emitted at one point and received at another.

There are three types of radiation associated with nuclear energy that ordinarily affect humans: alpha, beta, and gamma radiation. Alpha particles are the nuclei of helium atoms (two protons and two neutrons), and are readily stopped by even a thin material such as a sheet of paper. Beta particles (or rays) are high-speed electrons. Gamma radiation, like medical X rays, is composed of photons except that gamma radiation is emitted from the nucleus of atoms. Examples of each of these types of radiation encountered in everyday life are shown in Table 5-1 below.

The most meaningful units for measuring radiation dose to humans are the **rem and millirem** (1/1000 of a rem, abbreviated **mrem).** These units of measurement take into account the effect on living tissue (biological effectiveness) of the three types of radiation. Table 5-1 shows estimated average annual doses from some of the alpha, beta, and gamma sources to which individuals in the U.S. are exposed.

All three types of radiation (called "ionizing" radiation because of their effect on matter) are present in nature as well as through human activities. The natural radiation from the soil and water, and from cosmic radiation, is called "background radiation." There is an additional type of radiation present in nuclear power plants, that is, radiation due to neutrons. However, the only persons likely to be exposed even to low doses of neutrons are those working in the plant.

Chapter 8. Nuclear Energy

Coal-fired power plants emit measurable amounts of radiation due to the presence of naturally radioactive materials in the coal. The Geysers geothermal power plant in northern California emits naturally radioactive gases (radon) present in the steam issuing from the earth. In addition, we have all received a dose of about 4 mrem per year for the past 25 years from fallout from nuclear weapons testing. Finally, medical and dental X rays, while having clearly definable benefits, are the largest single man-made source of radiation exposure to the average U.S. individual.

Instruments are available for measuring the dose received by any individual, but because of their sensitivity and relatively high cost these are generally used only by persons routinely working with, and thus potentially exposed to radiation sources. The range of annual radiation doses for individuals is from 115 to 215 mrem without medical X rays, and the maximum could exceed 425 mrem in a year in which a gastrointestinal tract series of X rays were required.

Radioactive materials are produced in solid, liquid, and gaseous forms. **Solid wastes,** such as pieces of machinery contaminated with radioactive materials, used work gloves, and shoe covers, which have a very low level of radioactivity, are sealed in safe containers and shipped to licensed disposal areas for burial. Solid wastes with high levels of radioactivity, such as spent fuel. **Liquid wastes** include corrosion products, some fission products, and tritium. **Gaseous wastes** contain some gaseous fission products and also tritium. Gaseous and liquid wastes from nuclear power plants contain very low levels of radioactivity. These are carefully monitored according to U.S. Nuclear Regulatory Commission (NRC) regulations and are disposed of as discussed below.

Medium- and high-level radioactive liquid wastes are generated from fission products during fuel reprocessing. Radioactive corrosion products (e.g., iron, cobalt, zinc) result from chemical attack on radioactive metal parts in the reactor. The corrosion products are periodically removed from the reactor after being trapped from coolant in a liquid-waste processing stream.

In the fission process, uranium splits into two new (and lighter) atoms. These are called fission products. Many of the fission products are stable (nonradioactive), but others are highly radioactive. Only the radioactive atoms are hazardous. The fission products build up in the fuel during its use and are retained there until the fuel is removed from the reactor for reprocessing. Sometimes fission products escape from the fuel cladding, usually through tiny holes or cracks in the metal tube cladding, into the surrounding coolant, in which case they are collected as liquid or gaseous wastes, depending on the reactor type.

Most of the radioactive products pose no hazard because either they are produced in very small quantities or they have very short lifetimes and decay to innocuous species rapidly. A few radioactive products, because of their long lives, high yield, and sometimes their chemical properties, could pose a threat to the public if they were released to the environment in significant quantities.

Half-life is the time it takes for radioactivity to decrease to half its starting level. Each radioactive nuclide has a unique half-life, in the range from millionths of a second to billions of years.

Tritium is a radioactive form of hydrogen that can be produced in reactors in several ways. It can be removed from the system as a liquid or a gas. Because it can become part of water molecules and because of its long half-life (12.3 years), tritium is a potential biological hazard.

Xenon, krypton, and iodine are fission products that are normally retained entirely within the fuel elements but that occasionally escape into the reactor coolant, usually through tiny holes and cracks in the cladding. In this case they are removed from the system as gaseous wastes.

Radioactive wastes, separated from the reactor coolant as solids, are of very small volume and are sealed in special containers and shipped to licensed disposal areas. Low-level wastes are stored temporarily, permitting some of the radioactivity to decay, and then

Chapter 8. Nuclear Energy

diluted to harmless levels and discharged into lakes, rivers, and oceans.

Radioactivity in liquids discharged to the environment must be "as low as practicable" (ALAP) according to the NRC. The ALAP levels from light water reactors (LWRs) ensure that the exposure of an individual in an unrestricted area from discharged liquids is no more than 3 mrem per year.

Small amounts are discharged to the atmosphere in accordance with applicable regulations. However, Federal regulations require plants to store gases to permit further decay of the radioactivity before discharge. This markedly reduces the radioactivity of gases released.

The NRC also requires that gaseous radioactivity releases to the environment be ALAP. The gaseous ALAP levels from LWRs ensure that the exposure of a person sitting on a fence at the site boundary of a nuclear plant 24 hours a day, 365 days a year, will be less than 5 mrem; the average neighbor of the plant will receive less than 1 mrem annually. For comparison, everyone is exposed to natural background radiation of 100 to 150 mrem per year. Thus, the increased exposure to the average individual is less than 1% of the background exposure and it is lower than the variation of that background dose.

In 1970 the average annual dose from the nuclear industry was 0.01 mrem. Assuming nuclear energy becomes a dominant source of electricity, the average citizen will still receive a yearly dose of less than 1 mrem. Those people living near nuclear power plants will receive less than 5 mrem per year.

Radioactivity is a natural part of the coal as mined, and is emitted in the soot, fly ash, and smoke formed from burning the coal. The rate at which radioactivity is released by either nuclear or fossil (coal, oil and natural gas) power plants is negligible.

Significant levels of long-lived radioactive materials could build up in the environment from very low and dilute releases over long periods if not controlled. However, the U.S. Environmental Protection Agency has proposed regulations to protect the public by limiting the amounts of long-lived radioactivity" released to the environment from the entire nuclear fuel cycle (uranium milling, chemical conversion, enrichment, fabrication, power plant operation, reprocessing, and transportation). Also, the regulations would limit the annual dose to a member of the public to a maximum of 75 mrem to the thyroid or 25 mrem to the whole body or to any other organ.

It is impossible to have zero releases from nuclear power plants, just as it is impossible to have zero releases of pollutants from any power plant, or other processes for that matter. However, the releases from nuclear power plants are not only below the levels of significant environmental or human health effects, but they are made ALAP as discussed above. Control of releases and thus the control of the total environmental impact of nuclear power plants was an objective right from the start of nuclear power development rather than after the fact, as is commonly the case.

Yes, "biological concentration" can occur along the food chain. For example, dilute radioactive minerals from power plants can be taken up by algae, thus separating the mineral from the water and concentrating the radioactivity in the process. The algae are then consumed by zooplankton and the zooplankton by small fish in turn, being further concentrated at each step. The chain continues through larger fish to birds and their eggs, to soil, to groundwater, to vegetables, and meat animals to humans.

Radionuclides

Certain types of atoms disintegrate spontaneously through time; radiation is the result of this atomic disintegration.   The atoms emit an invisible radiation capable of passing through solid objects. The radiation from uranium minerals fall into three types, according to the direction of deflection in a magnetic field. These three types of radiation were called

> alpha (α),

> beta (β), and

gamma (γ) radiation.

Gamma radiation is the equivalent of x-rays. We require a basic understanding of alpha, beta, and gamma emissions and emitters and understanding of the effect of neutrons.

Dangerous Radioactive Products

Product	Type of Radiation	Half-life
Krypton-85 (Kr85)	beta and gamma	10 years
Strontium-90 (Sr90)	beta	20 years
Iodine-131 (I131)	beta and gamma	8 days
Cesium-137 (Cs137)	beta and gamma	30 years
Tritium (Tl)	beta	12 years
Cobalt-60 (Co60)	beta and gamma	5 years
Carbon-14 (C14)	beta	5770 years

Often, the original radionuclide, called the parent, decays to a nucleus that is also unstable, called the daughter; the daughter often decays even further. Chains are not uncommon where radioactive daughters produce radioactive second daughters, which in turn produce radioactive third daughters and so on. For example, for the decay chain beginning with U233 we observe ten steps before the stable Pb207 is reached. Radioactive decay is commonly described in terms of the **half-life**, or the time for one half of the radioactive atoms to have disintegrated.

Ionizing Radiation

An emitted particle always has a certain amount of kinetic energy. This energy is lost in collisions with target atoms as the alpha, beta, and gamma radiations pass through different materials. If the material is human tissue, the energy gained (by the tissue) causes disordering in the chemical or biological structure of the tissue and may produce cell deaths or subsequent dysfunction of the cells. The damage to human tissue is directly related to the amount of energy deposited in the tissue by the alpha, beta, and gamma particles. This energy, in the form of ionization and excitation of molecules, results in heat damage to the tissue or even radiation burn.

Alpha radiation, emitted from the nuclei of certain radioactive atoms, is made up of charged particles, each traveling at approximately 10,000 miles per second. Each particle is the same as the nucleus of a helium atom, that is, each particle consists of 2 protons and 2 neutrons. The alpha particles interact with other atoms. The electrons of these other atoms receive a portion of the energy of the alpha particles; the result is the production of an ion pair; that is, a negative electron with an associated positive ion. These interactions result in a rapid depletion of the alpha energy, so the range of the alpha particle is only 1 to 8 cm in air. Even the strongest alpha particles are stopped by the epidermal layer of the skin and rarely reach the sensitive layers. If such a particle runs into a solid object, notably human skin cells, the energy is rapidly dissipated. Thus alpha particles present no direct problem of external radiation damage to humans. However, alpha radiation can cause health problems when emitted inside the body where protective layers are not present to diffuse the energy. Humans are typically contaminated with material that emits alpha particles only if the material is inhaled, ingested, or absorbed in a skin wound.

Beta radiation is made up of electrons emitted from the nucleus of a radioactive atom at a velocity approaching the speed of light. The interactions between beta particles and the atoms of pass-through materials are much less frequent than alpha particle interactions. The resulting slower rate of energy loss enables beta particles to travel several meters in air and several centimeters through human tissue. Exposed organs such as eyes are sensitive to beta damage. Internal organs are generally protected by the skin.

Gamma radiation produces rays that are like medical x-rays in that they are composed of photons. These invisible, electromagnetic rays are emitted from the nucleus of radioactive atoms. Because of their neutral charge, gamma photons collide randomly with the atoms of the material as they pass through. A typical gamma ray has a unique relaxation length for different pass-through materials: for example, lead, water, and air have relaxation lengths of 5, 50, and 10,000 cm respectively. The dose of gamma radiation received by unprotected human tissue can be significant because the dose is not greatly impacted by air molecules.

Radiation doses by Canadian Nuclear Safety Commission

What is a radiation dose?

When ionizing radiation penetrates the human body or an object, it deposits energy. The energy absorbed from exposure to radiation is called a dose. Radiation dose quantities are described in three ways: **absorbed**, **equivalent**, and **effective**.

Dose quantities

Absorbed dose, energy deposited in a kilogram of a substance by radiation. Equivalent dose, absorbed dose weighted for the degree of the effect of different radiations (radiation weighting factor wr). Effective dose, equivalent dose weighted for susceptibility to effect of different tissues (tissue weighting factor wt).

Absorbed dose

The amount of energy deposited in a substance (e.g., human tissue), is called the **absorbed dose.** The absorbed dose is measured in a unit called the gray (Gy). **A dose of one gray** is equivalent to a unit of energy (joule) deposited in a kilogram of a substance.

Equivalent dose

When radiation is absorbed in living matter, a biological effect may be observed. However, equal absorbed doses will not necessarily produce equal biological effects. The effect depends on the type of radiation (e.g., alpha, beta, gamma, etc) and the tissue or organ receiving the radiation. For example, 1 Gy of alpha radiation is more harmful to tissue than 1 Gy of beta radiation.

A radiation weighting factor (w_R) is used to equate different types of radiation with different biological effectiveness. This weighted absorbed quantity is called the equivalent dose and is expressed in a measure called the sievert (Sv). This means that 1 Sv of alpha radiation will have the same biological effect as 1 Sv of beta radiation.

Because doses to workers and the public are so low, most reporting and dose measurements use the terms millisievert (mSv) and microsievert (μSv) which are 1/1000 and 1/1000000 of a sievert respectively. These smaller units of the sievert are more convenient to use in occupational and public settings.

To obtain the equivalent dose, the absorbed dose is multiplied by a specified radiation weighting factor (w_R). The equivalent dose provides a single unit which accounts for the degree of harm of different types of radiation.

Dose from background radiation

Radiation has always been present all around us. In fact, life has evolved in a world containing significant levels of ionizing radiation. It comes from space, the ground, and even within our own bodies. The doses due to natural background radiation vary depending on location and habits.

Dose from cosmic radiation

Regions at higher altitudes receive more cosmic radiation. According to a study by Health Canada, the annual effective dose of radiation from cosmic rays in Vancouver, British Columbia, which is at sea level, is about 0.30 mSv. This compares to the top of Mount Lorne, Yukon, where at 2,000 m, a person would receive an annual dose of about 0.84 mSv. Air travel also increases exposure to more cosmic radiation, for a further average dose of 0.01 mSv per Canadian per year.

Dose from terrestrial radiation

Chapter 8. Nuclear Energy

There are also natural sources of radiation in the ground. For example, some regions receive more terrestrial radiation from soils that are enriched with uranium. The average effective dose from the radiation emitted from the soil (and the construction materials that come from the ground) is approximately 0.5 mSv a year. However, the dose varies depending on location and geology, with doses reaching as high as 260 mSv in Northern Iran or 90 mSv in Nigeria. In Canada, the estimated highest annual dose for terrestrial radiation is approximately 1.4 mSv measured in the Northwest Territories.

Dose from inhalation

The earth's crust also contributes to our levels of exposure. Radon gas, which is produced by the earth, is present in the air we breathe. Radon gas naturally disperses as it enters the atmosphere from the ground. However, when radon gas enters a building (through the floor from the ground), a concentration of it tends to build up. Long-term exposure to elevated levels of radon increases the risk of developing lung cancer. The worldwide average annual effective dose of radon radiation is approximately 1.2 mSv.

Dose from ingestion

A number of sources of natural radiation that penetrate our bodies through the food we eat, the air we breathe and the water we drink. Potassium-40 is the main source of internal irradiation (aside from radon decay) found in a variety of everyday foods. The average effective dose from these sources is approximately 0.3 mSv a year.

Effects of Radiation

There are two kinds of biological effects of radiation: somatic and genetic effects.

> **Somatic effects** are effects on the exposed individual and include a slightly increased incidence of cancer and life shortening. At the low levels of concern with respect to nuclear power production and utilization, the somatic effects are negligible.

Genetic effects are effects that are transmitted to the offspring of the exposed individual by mutations in the genes.

The biological effects of radiation are better known than those of any other hazard. But the biological effects of radiation at extremely low doses and dose rates are not well known and can only be estimated from data at higher levels. The data at low levels are not yet sufficient in numbers and population generations, even using test animals, to be definitive or conclusive. Thus, to be on the safe side in the use of radiation and radioactive materials, maximum possible effects are used for setting standards. It has been **assumed** that biological effects are directly proportional to the total dose received regardless of how low the dose goes. That is, there is no threshold level below which there are no effects. This is called the "linear hypothesis" of radiation effects. Thus, for example, if 100,000 mrem of radiation dose to a large population were to produce 10 observable effects, then 10,000 mrem will produce 1.0 effect, and 1000 mrem will produce 0.10 effect and so on down to the lowest dose considered.

The **lifetime** dose of the average American from the nuclear power industry is less than 1 mrem (0.01 mrem per year for 70 years) or less than 1% of his **annual** dose from natural background radiation.

Radiation can cause mutations—abrupt, inheritable changes—in living cells. If radioactivity releases from all nuclear plants in the U.S. were at the current **maximum permissible** levels, it has been estimated that 24 genetic mutations per year would occur in the U.S. population. But the rate from the **actual average dose** (0.01 mrem per year) would be less than one mutation in four years. Since the spontaneous genetic mutation rate is 800,000 per year, the operation of nuclear power plants would increase this figure by less than 0.24 to 800,000.24.

The incidence of lung cancer among uranium miners, especially those who smoke cigarettes, has been higher than that of the general public. Uranium miners, primarily those employed in subsurface mines, are exposed to radon, a radioactive gas, and other radioactive

Chapter 8. Nuclear Energy 195

decay products of uranium and thorium. Initially in the 1940's and 1950's only general protection was provided for the miners. In 1971 stricter regulations were applied, limiting the exposure of miners to four working-level-months per year, and requiring protective respiratory equipment for the miners. Better ventilation of the mines has eliminated the problem.

Residents near the large uranium mill tailing piles may get lung cancer, like the uranium miners, from exposure to radon gas and its radioactive decay products. The mill tailings are huge piles of earth (e.g., two 230 feet high piles in Colorado cover 146 acres) left after the economically recoverable uranium has been extracted from the ore. The decay products of uranium formed prior to its extraction from the ore along with the residual uranium remaining in the tailings continue to be a source of radon gas for thousands of years. The radon escapes into the atmosphere from these piles, even though they are covered with up to 20 feet of ordinary soil, and can result in significant doses to nearby residents. The radon release from the mill tailing piles is a negligible fraction of that naturally released from uranium that is widely distributed in small amounts in the upper 10 feet of the earth's crust. Thus, radon from the mill tailing piles is not a general problem but is an important and unsolved local problem. The states of Wyoming Utah, New Mexico, and Colorado, where most of the mining and milling is done, and the U.S. Environmental Protection Agency and the U.S. Nuclear Regulatory Commission have implemented programs to solve the problems.

Biological effects

Ionizing radiation is given off by decaying radioisotopes, or radionuclides, typically as alpha or beta particles or gamma rays. Ionizing radiation can penetrate a human cell and then ionize (i.e., knock an electron loose from) a cellular chemical, altering molecules important for normal functioning. When this happens (and it happens frequently, because of natural or "background" radiation), cells usually repair themselves. But occasionally they die or are transformed. Other electromagnetic radiation, such as microwaves or light, may also injure cells but does not cause ionization.

Although it is understood how radiation directly affects the human body, it is not known how much low radiation exposure is required to produce subsequent cancer and other illness. This lack of knowledge gives rise to controversy over what level of radiation exposure is acceptable. Biological effects from radiation can include damage to the circulatory system, carcinogenesis, and decrease in organ function because of cell killing. Genetic effects are the result of ionizing radiation damages the genetic material of the cell. Genetic effects do not appear in the individual receiving the radiation but appear in that individual's descendants.

The health effects of a given dose of radiation depend on a large number of factors, including

1. magnitude of the absorbed dose

2. type of radiation

3. penetrating power of the radiation

4. sensitivity of the receiving cells and organs rate at which the dose is delivered

5. proportion of the cell/organ/human body exposed

Ionizing radiation injures cells in two ways, one immediate and the other delayed. High-level doses—a term used loosely for exposures over 100 rem—inevitably produce immediate, direct effects: skin burns, hair loss, bone marrow destruction, and damage to the intestinal lining. But the effects of low doses—a loose term for exposures under 10 rem—are not immediately apparent because they involve the cancerous transformation of cells. Since low doses produce effects (cancer, genetic effects) that have very long latency periods, good data does not exist for low doses. The process that produces cancerous tumors may take place in some persons and not in others. Although the relation between very low doses of radiation

Chapter 8. Nuclear Energy

and cancer is still unresolved, there is a consensus about the effects of high and moderately high doses of radiation. The direct, immediate effects of radiation are seen in all cases, from the highest doses down to about 100 rem. Persons exposed to levels of 10-50 rem have an increased likelihood of developing cancer.

Radiation sickness (circulatory system breakdown, nausea, hair loss) and resulting death are acute somatic effects occurring after very high exposure as from a nuclear bomb or intense radiation therapy. The accident at the Chernobyl nuclear generating plant in 1986 resulted in about 30 deaths from acute effects during the 2 months after the event. These individuals received estimated gamma-radiation wholebody doses of approximately 400 rads. Lower doses, but above 100 rads, will lead to vomiting, diarrhea, and nausea in humans.

To protect public health, one should act to minimize unnecessary radioactive exposure. Since radiation is a carcinogen, there is assumed to be no threshold for damage. As effects of exposure have been expressed over the years since the first atomic bomb, risk estimates have become more refined. The number of cancer fatalities in the United States per year per 10^6 persons is 1930. Exposure to an additional rad per year would increase the number of fatal cancers per million of population by about 15 percent.

Natural radiation

Radon gas (Rn222) is a daughter of U233, which is ubiquitous in soil, rock, building materials, and so forth. Since Rn222 is chemically an inert gas, it is not trapped chemically. It is present not only in uranium ore but also in ordinary soil and in a number of rock formations in the United States, most notably in the Piedmont plateau and other parts of Appalachia. Rn222 from soil can accumulate in the basements of buildings and in occupied areas of buildings if ventilation is not adequate. Rn222 itself has a short half-life (3.4 days) and, being inert, is not absorbed or trapped in the lung. During its transit through the human lung, however, it can decay to Po218 and eventually to Po214. These nuclides are adsorbed to the lung surface. The National Academy of Sciences now estimates that the annual average dose from inhaled alpha emitters is 2500 mrem per year, and that this dose is almost entirely due to the radio progeny Po218 and Po214. Rn222 is probably the single greatest "background" source of human exposure.

Two regions of high natural radiation are in Africa and in Brazil. In Gabon, West Africa, near a place called Oklo, the concentration of U-235 was so high about 1.8 billion years ago that a natural chain reaction took place. This natural nuclear fission reactor operated for a period of about one million years. Tons of uranium underwent fission. Plutonium and other transuranic substances were produced. Even though this natural reactor area has been subject to rainfall and other weathering agents, the plutonium and fission isotopes have migrated only a few meters from the place of their production.

The other region is Morro do Ferro in Brazil, where there is a weathered mound 250 meters high formed of from ore body containing an estimated 30,000 metric tons of thorium and 100,000 metric tons of rare earths. The radiation level is one to two mRoentgens per hour over an area of 30,000 square meters. The mound supports both animal and plant life. Photographs show plants truly glowing in the dark because of the high absorbed radioactivity in the vegetation. A colony of rats living in burrows in a mound breathe air containing radon at levels up to 100,000 pCi per liter. Fourteen rats were trapped and autopsied, and no abnormalities were found. Yet the radiation dose to the rats' bronchial epithelium was estimated to be between 3,000 and 30,000 rems per year, roughly three times the concentration that should produce tumors or other radiation effects. Apparently, life —plant, animal, and human—can exist in places where the background natural radiation is abnormally high, even with exposures exceeding 10,000 millirems per year, the generally accepted cutoff between low-level and high-level radiation.

Chernobyl

Chapter 8. Nuclear Energy

On April 26, 1986 there was a devastating explosion and fire at the Chernobyl nuclear power plant. Important parts of the story have not previously been known, or their significance fully appreciated, outside of Russia and Ukraine. The picture is a sobering one. Conditions at the plant were far more uncontrolled than had been thought, and measures taken to quell the nuclear fire were almost totally ineffective. The only good news in the story is that apparently the nuclear reaction eventually put itself out.

By the afternoon of April 27, 1986 - the day after the accident - in a desperate bid to do something about the release of radiation from the plant and the many fires still raging inside it, orders were given to begin dumping various materials onto the reactor core, both to contain it and stop the burning of graphite in the core itself. By the end of day six, helicopters had dropped more than 5,000 tons of sand, clay, lead, boron (a neutron-absorbing element) and dolomite, a mineral that releases carbon dioxide gas when heated. Consultants thought the gas might help extinguish the fires.

The pilots hit the target they had been told to aim for: a "red glow" in the reactor building. It is now known that the red glow was not the reactor core, but an unknown mass about 50 feet from the nuclear reactor vessel. Nobody is sure what the glow actually was, and the answer may not come until someone manages to drill into the four-story-high mound of dropped material that still sits there today too radioactive and unstable to be studied. The glow may have been a small chunk of the reactor core, blown into that area by the force of the initial explosion that shattered the reactor.

It is now clear that virtually none of the dropped material entered the reactor vessel, where most of the radioactive core burned and melted unchecked for nine days, deep in the wrecked building and hidden from the helicopter pilots' view. Until now the original conclusion voiced by Russian officials in an international meeting in Vienna in 1986 has been generally accepted: That the pilots extinguished the burning graphite, smothered the nuclear reaction, and halted the massive outpouring of radioactive material.

Some specialists have known since 1991 that the dumping strategy helped little if at all, but it had not been revealed before that essentially none of the material reached the core. Russian scientists began to suspect that the helicopter pilots' heroic flights were in vain as soon as they saw that the core vessel was empty; it should have contained mounds of the material. But it took detailed analyses carried out by Russian and Ukrainian scientists, and first brought to light to the public in 1988 to confirm that suspicion. A chemical analysis of samples of the lava-like core material recovered from inside the sealed reactor building showed it contained no lead, no boron - none of the material that should have been incorporated into the molten core if it had indeed landed on the overheated nuclear fuel.

That conclusion may not be accepted by all researchers. Morris Rosen, assistant director of the International Atomic Energy Agency's division of nuclear safety, remains skeptical. He said last week: "I can vouch for the fact, since I flew over" the reactor in the days after the explosion, "that the reactor eventually stopped releasing radiation.   Material certainly got into the core region, I can assure you." If that wasn't what stopped the releases, then what did?

It was on May 3, 1988, that the official version of what happened began to come unraveled. That was when engineers succeeded in drilling a shaft through a series of thick concrete walls and rubble into the core of the wrecked reactor, hoping to determine the condition of what remained of the radioactive fuel and its graphite shell. Their startling discovery, made with a remote camera guided through the shaft and into the reactor vessel, began a long and laborious process of piecing together the sequence of events in the ten crucial days that followed the devastating explosion. "We saw that there was no graphite, no core, nothing. Empty. We were extremely frightened by this. We didn't know where the fuel was," recalled Alexander Borovoi, head of the nuclear and radiation safety department at Moscow's Kurchatov Institute and research director at the Chernobyl site.

Chapter 8. Nuclear Energy

To learn what had become of the missing core, where more than 100 tons of uranium fuel had been, the scientists worked during the ensuing days to insert cameras into shafts they drilled into the lower reaches of the reactor building. There, a few days later, they began to find parts of what was left of the core. It was the first piece in the puzzle of reconstructing the ten-day period when the core was in a super-hot, uncontrolled reaction. "We realized that the core had melted and flowed into the lower parts of the building," Borovoi said.

It turned out that the reactor had undergone what is considered a worst-case scenario for a nuclear accident: a complete core meltdown. Now, using information gathered by scientists who have worked at the site ever since the disaster, a reconstruction of the ten-day sequence of events reveals what really happened.

Researchers have previously documented the fundamental causes of the world's worst nuclear accident a combination of the plant's inherently risky design and a series of inexcusable mistakes by its operators caused the initial explosion, and changes have been made to the remaining reactors of the same RBMK type to make them safer. But what has not been fully understood before was the total failure of virtually all the measures taken after the accident. If another nuclear accident should occur - something that many analysts consider likely even after the fixes that have been made to the Russian -designed RBMK reactors - it is important to understand why the countermeasures failed.

It took many perilous human forays into the plant, many drilled holes, exploration with remote cameras and a painstaking analysis of 160 samples from the solidified remains of the molten core, but scientists feel that they now understand most of what really happened during those 10 days of the "active phase" of the nuclear disaster. For almost nine days, the graphite blazed and the molten core material slowly ate its way down through a six-foot radiation shield. When it finally melted all the way through, on day nine, the molten core dropped through into the lower level of the plant. There, the hot liquid quickly spread out over concrete floors, found its way down through steam shafts and stainless steel tubing, and eventually spread out so much that the uncontrolled nuclear reaction stopped. The mass then cooled rapidly and "froze" in place. It froze solid in what appears to be a matter of minutes, judging from photographs taken inside the reactor that show, for example, core material spouting from a steam fitting in a kind of waterfall effect. It can be seen that it wasn't oozing out like toothpaste, it was running out and froze in place. There it sits today, solidified in midstream like a flow of lava on the flanks of a volcano - except that it is still emitting deadly radiation.

The fact the core remained uncovered throughout the active phase helps explain why far more radioactivity escaped than was originally estimated by the Russian scientists. The total release was four to five times more than the Russia claimed in 1986. For example, the Russians said about 10 percent of the radioactive tellurium in the core, or about 12 million curies worth, was dispersed. Independent analysis by the US Nuclear Regulatory Commission, based on measurements of the radioactive plume picked up far from the site, estimated a total release of 7 percent. But the new estimate are that at least 25 percent of the tellurium, a radioactive isotope of the element that has a half-life of a little over three days, was dispersed - or at least 30 million curies. That big a release could help explain some health effects observed in the region around the plant, including a small apparent increase in the incidence of childhood thyroid cancers and an almost tenfold increase in childhood anemia.

While contamination in most of the surrounding area has since been quite effectively cleaned up, some scientists fear that the poor condition of the reactor building's protective housing could lead to further exposure of people in nearby towns outside the 18 mile exclusion zone. The so-called "sarcophagus" built to seal in the reactor building, preventing further escape of nuclear waste, has been effective so far, but its long-term ability to contain the 180 tons of radioactive material is much in doubt. The hastily-built foundations erected on what was once marshland, have already begun to shift, and the building itself is riddled with holes. Western engineers were asked to suggest ways to shore up and seal this

Chapter 8. Nuclear Energy

building, but the proposals are all expensive and neither Ukraine nor Russia has the money, so they are appealing for help from other nations.

Thanks to the efforts of fast-footed and courageous scientists who explored the interior of the entombed reactor and took photographs and samples of the reactor's molten remains, the world now has the most complete documentation to date of the sequence of events inside the out-of-control reactor. And thanks to the efforts of scientists at the scene such as Russia's Borovoi, much is now known of the precarious state of affairs at the site today, and of the urgent need for measures to prevent further fallout from the disaster.

Fukishima

The Fukishima disaster in Japan was treated in Chapter 1. It was a triple nuclear meltdown Many blame Fukishima for the collapse of fisheries. It is claimed that the radiation flowing from the plant is increasing. The half-life of the various radioactive elements ensures that the radiation levels will just keep rising. It is said that Fukishima is an 'extinction level event' of the Pacific; it will be felt by every living thing. Radiation works slowly but surely to destroy everything it comes into contact with, The Japanese government has publically declared that the decommissioning could take around four decades and cost around $189 billion. In fact, those on the ground are faced with insurmountable problems. There is no precedent for this kind of situation. Robots and cameras have already provided valuable pictures, but it is still unclear what is really going on inside. In addition to the damages already incurred, there are many factors that could exacerbate the situation. For example, another earthquake could strike the area and cause radiation explosions. Earthquakes are not unusual in Japan. If the reactors encounter a big earth tremor, they could be destroyed and scatter the remaining nuclear fuel and its debris, making the Tokyo metropolitan area uninhabitable. Local sources indicate that deaths related to the accident rarely go reported, especially because most of these people die outside of the workplace.

Radioactive Waste

There are a number of sources of radioactive waste: the nuclear fuel cycle, radiopharmaceutical manufacture and use, biomedical research and applications, and a number of industrial uses. The behavior of radionuclides is determined by their physical and chemical properties; radionuclides may exist as gases, liquids, or solids and may be soluble or insoluble in water or other solvents.

Vast inventories of radioactive wastes require disposal. These wastes consist of **uranium mill tailings**, **transuranic wastes** (defined as materials containing more than 100 nanocuries of transuranic material per gram of waste), **high-level wastes** (which include spent fuel rods and the sludges left behind when spent fuel rods are reprocessed to extract plutonium), and **low-level wastes** (which include those wastes not included in the other categories).

These designations follows the U.S. Nuclear Regulatory Commission (NRC) which classifies radioactive wastes into the following categories: High-level waste (HLW), Transuranic (TRU) waste, By-product material, Uranium mining and mill tailings, and Low-level waste (LLW.

There are various options for storing nuclear waste. The problem with nuclear waste is that it is dangerous. Nuclear waste can emit radiation for centuries. It could potentially become unstable if handled and stored improperly, setting off a chain reaction by accident. If it fell into the wrong hands, it could be used to make a dirty bomb which would spread radiation over an inhabited area. Nuclear waste storage focuses on finding safe and secure ways to store spent nuclear fuel and other forms of nuclear waste until they stabilize enough that they do not pose a threat to humans, wildlife, and the environment.

Temporary nuclear waste storage is typically the first step. In many cases, nuclear waste is extremely hot when it is generated, and it needs some time to cool down. At nuclear power plants, spent nuclear fuel is submerged in pools filled with boric acid to allow it to cool and stabilize. These pools are usually made from steel-lined

Chapter 8. Nuclear Energy

concrete to prevent leakage, and they are definitely a temporary measure.

Once nuclear waste is cooled, it can be moved into dry cask nuclear waste storage. Such storage involves extremely durable barrels which are designed to prevent leaks of radiation. The casks can be filled with waste and stored above ground safely, although they are also designed ultimately for temporary storage. Over the thousands of years which may be required for the waste to stabilize, the casks could fail or be breached.

For long term nuclear waste storage, it is necessary to find a safe place to keep the material while it breaks down. Burial is one method which has been widely promoted, as the nuclear waste can safely break down underground or under the ocean floor in remote areas. The issue with burial is that the nuclear waste could leak or be breached by earthquakes or human activity. There are also concerns about the fact that in several thousand years, it is unlikely that knowledge of the site as a dangerous location will survive, which means that future civilizations could unwittingly release toxic materials into the environment when they breach storage facilities.

Nuclear energy production produces radioactive waste at every stage. Mining and milling generate the same sort of waste that any mining and milling operations generate except that this waste is radioactive. Mining and milling dust must be stabilized to keep it from dispersing, and leachate must be prevented from contaminating waterways and groundwater. Partially refined uranium ore (called "yellowcake" because of its bright yellow color) is then enriched in the fissile isotope U235, and nuclear fuel is fabricated. Enrichment and fabrication produce TRU waste. he nuclear fuel is then inserted into a nuclear reactor core where a controlled fission reaction produces heat, which in turn produces pressurized steam for electric power generation. The steam system, turbines, and generators in a nuclear power plant are essentially the same as those in any thermal electric power plant (like a coal plant).

In the United States, plutonium for weapons is produced by irradiating U238 with neutrons in military breeder reactors. The plutonium, along with uranium and neptunium, is then extracted by dissolving the entire irradiated fuel element in nitric acid and then extracting with tributyl phosphate. Further partition and selective precipitation result in recovery of plutonium, uranium, neptunium, strontium, and cesium. The fissile isotopes of plutonium and uranium are categorized as special nuclear material; the remainder of these is considered byproduct material. Both the acid solvent, which is ultimately neutralized, and the organic extraction solvents contain high concentrations of radioisotopes and are classified as HLW. This process also yields TRU and low-level wastes.

The only use for plutonium in the United States is in weapons manufacture; commercial spent fuel is not reprocessed. In France, however, plutonium is produced in breeder reactors for use in nuclear electric power generation.

The primary and secondary coolants in a nuclear generating plant pick up considerable radioactive contamination through controlled leaks. Contaminants are removed from the cooling water by ion-exchange columns. The loaded columns are Class C low-level radioactive waste. Class A and B wastes are also produced in routine cleanup activities in nuclear reactors. Eventually, the reactor core and the structures immediately surrounding it become very radioactive, primarily by neutron activation, and the reactor must be shut down and decommissioned. Although the two small reactors now being decommissioned are being treated like LLW, the ultimate implications of decommissioning are not yet clear.

A tenfold increase in fatal lung cancers among uranium miners has been established resulting in an extremely high risk of lung cancer among uranium miners. The extent of the dangers associated with other aspects of the nuclear fuel cycle—milling uranium ore, fabricating fuel rods, transporting fuel rod assemblies, and other operations—indicate significant health effects from these activities. Considerable evidence shows that nuclear workers experience elevated risks from low dose external radiation for leukemia,

Chapter 8. Nuclear Energy

lymphatic, and hematopoietic cancer combined, and to a lesser extent for solid tumors and for all cancers combined. Low-level radiation emitted during the normal operation of nuclear power generating facilities may well cause increased cancer rates and deaths in the general population. A recent study of leukemia incidence around the Pilgrim nuclear power station in Plymouth, Massachusetts, raises serious questions about the safety of reactors during routine operations.

There is always the possibility of a reactor core meltdown accompanied by a catastrophic release of radiation. This possibility increases because of the effects of aging with the attendant deterioration due to heat, corrosive chemicals, and mechanical stress. Also, nuclear reactors are subjected to the degrading effects of intense neutron irradiation, which makes the steel brittle and more likely to fracture. This is particularly true for older reactors. The 1986 accident at the Chernobyl nuclear power plant demonstrated that catastrophic reactor accidents were not merely theoretical possibilities. Millions of curies of radionuclides—including I-131, Sr-90, Ce-137, and Pu-239—were released into the surrounding countryside, contaminating 5.5 million hectares and exposing 2.5 million people to elevated radiation. Hundreds of workers and others in the immediate vicinity received lethal doses of radiation, and many died from acute radiation sickness. The Chernobyl accident deposited significant radioactive fallout over portions of Eastern and Central Europe, and in areas as far away as Sweden, Italy, and Wales. Agricultural and dairy products in many localities had to be destroyed.

High-level waste (HLW)

HLW includes two categories: spent nuclear fuel from nuclear reactors and the solid and liquid wastes from reprocessing spent fuel to recover plutonium and fissile uranium. The commercial nuclear power industry in the United States alone will have produced 41,000 metric tons of high-level wastes by the year 2000. There is no established safe technology for dealing with high-level wastes. In the United States, spent fuel rods are currently stored in cooling pools at reactor sites pending creation of an acceptable repository for them. No reported leaks from these storage pools have been reported, but there have been massive leaks from storage facilities for high-level waste from the nuclear weapons program.

Yucca Mountain in Nevada was proposed as a site for a nuclear waste repository. For many years, the Department of Energy has been studying the bald ridge of Yucca Mountain in Nevada. Their purpose is to determine whether Yucca Mountain will make a suitable burial ground for spent radioactive fuel currently accumulating in overcrowded facilities at nuclear power plants around the country. Yucca Mountain is located northwest of Las Vegas. There are 106 miles from Las Vegas to Yucca and 130 miles (209 kilometers) by car. Its history by year follows.

1982. Congress passes Nuclear Waste Policy Act, requiring establishment of a place to store nuclear waste. The act requires 2 repositories -- one on each side of the country.

1983. 9 locations in 6 states are selected for review.

1984. DOE issues site suitability guidelines

1986. DOE nominates 5 sites. 3 sites are then selected for further investigation -- Hanford, Washington; Deaf County, Texas; and Yucca Mountain. Yucca Mountain was chosen based on several factors: distance from a major population center, desert location, in a closed hydrologic basin, surrounded by federal land and protected by natural geological barriers.

1987. Congress amends the 1982 legislation, stopping the selection process. Yucca Mountain becomes the selected site.

1988. DOE holds public hearings on site characterization.

1993. DOE begins grading work on first phase of the Exploratory Studies Facility.

Chapter 8. Nuclear Energy

1994. Tunneling into Yucca Mountain begins. Critics say the portal ramps and entrance are constructed for use as a repository and not a study area.

1995. Tunnel boring machines encounter loose ground at various points.

1996. Testing reveals that Yucca Mountain may not be the best site to store nuclear waste.

1998. DOE fails to meet January deadline for waste acceptance. Lawsuits are filed by the state and nuclear industry. Proposed legislation for temporary storage dies in Congress. DOE released Yucca Mountain Viability Assessment report which declares the site as viable but says much work must be done before it can be officially recommended.

2000. New site suitability guidelines are issued. The new guidelines do not include many requirements of the original Nuclear Waste Police Act. The state of Nevada files a lawsuit in an effort to stop further development. The site characterization work continues.

2001. EPA announces proposed radiation standards for Yucca Mountain. The state of Nevada files a lawsuit against the EPA, arguing the standards are inadequate. DOE is forced to investigate reports of collusion between itself, its contractors and the nuclear industry.

2002. Yucca Mountain is officially recommended in February. Nevada issues a Notice of Objection in April. Both the U.S. House of Representatives and the U.S. Senate override the objection in July. On July 23, a resolution is passed that allows the DOE to take the next step in establishing a repository.

2003. Nuclear Waste Technical Review Board releases a report expressing concerns about Yucca Mountain. The board recommends additional research.

2004. The D.C. Circuit of Appeals decides that the EPA's 10,000-year compliance period for radiation protection at Yucca Mountain is illegal. The court rules that the EPA must reissue its compliance period rule. The NRC must also reissue its licensing rule, which is based on the EPA's regulations. In November, the DOE announces it will not submit its Yucca Mountain license application to the NRC in December as planned. It is announced that a revised schedule will be released in 2005.

2005. In January, the DOE unveils its plan for above ground nuclear waste storage. They announce plans to ship nuclear waste to Yucca Mountain in "dedicated trains" that would cross multiple states. In March, the DOE accuses the US Geological Survey of falsifying QA documentation.

2006. DOE says it will submit licensing application to NRC in June. They set a new target date of 2017 for the opening of Yucca Mountain as a repository.

2007. DOE issues environmental studies. Certifies license application document database is complete. The Walker River Paiute Tribe withdraws its permission to allow nuclear waste to be shipped through its reservation. In December, the NRC rejects Nevada's challenge to the database. Congress cuts Yucca Mountain's budget to $390 million in 2008.

2008. DOE submits its license application to NRC for approval. Nevada files a petition urging the NRC to reject the application.

2009. Secretary of Energy says that Yucca Mountain is not a workable option and the DOE terminates its efforts to obtain a license.

2010. A 15-member panel of experts is named to look into ways to handle nuclear waste. It became known as the Blue Ribbon Commission on America's Nuclear Future. They were told to issue a report in 18 months. In February, the NRC judges halt Yucca license hearings. In March, the group asks Congress to keep Yucca Mountain alive. In May, the state of Nevada files a motion with the NRC, asking

Chapter 8. Nuclear Energy

it to approve the DOE's application to pull out of Yucca Mountain. In June, the NRC says that the DOE cannot withdraw.

2009-2012. House lawmakers attempt to insert funding for Yucca Mountain into the DOE spending bill. However, the Senate rejects it each year.

2011. Nevada is in a state of limbo at the beginning of the year. In July, Washington state and South Caroline file lawsuit to compel the NRC to resume its consideration of a nuclear waste repository at Yucca Mountain. A 3-judge panel rules the lawsuit is premature until the NRC makes a final decision about Yucca Mountain. In September, the NRC allows the continuation of plans to close the controversial site.

2012. The Blue Ribbon Commission recommends that Congress create and fund a new organization dedicated solely to managing spent nuclear fuel.

2012. U.S. Senate Energy and Water Appropriations Committee produce bill that takes first step to address nation's nuclear waste problem. The bill seeks to establish a more cooperative approach for the government to recruit states and communities to host temporary nuclear waste storage sites and a permanent repository.

2013. DOE issues a strategy for managing spent nuclear fuel in response to the commission's recommendations. The plan calls for an interim storage facility to be established by 2021. U.S. Court of Appeals for District of Columbia tells NRC to restart Yucca Mountain licensing proceeding using appropriated funds even though there is not enough money. The NRC orders the process to restart.

2014. The NRC releases part of its report on suitability on Yucca Mountain as a disposal spot.

2015. The NRC releases more of its report. The safety evaluation report includes the staff's recommendation that the commission should not authorize construction of the repository because the DOE has not met certain land and water rights requirements.

2016. Work has stopped on the exploratory tunnel and it has been boarded up. The site has basically been abandoned. The status of the project is uncertain. Spent nuclear fuel is currently being stored at 121 sites across 39 states. More than 160 million Americans live within 75 miles of the sites. No one lives within 5 miles of Yucca Mountain and very few people live within 15 miles.

2017. The government calls for a restart of licensing for Yucca Mountain. $120 million is included in budged blueprint for fiscal year 2018, which began Oct. 1, 2017. Bill to license and expedite the licensing and development of the nuclear waste site is passed by the subcommittee. Nevada reiterates his pledge that it will oppose any attempt to continue development of Yucca Mountain.

2018. The House of Representatives approves bill to revive Yucca Mountain on May 10. The bill directs the DOE to continue its licensing process for Yucca Mountain while moving forward with a separate plan for a temporary storage site in New Mexico or Texas.

The long-term desire is that fuel rods from commercial reactors will ultimately be stored in deep geological repositories, with the intention that the waste is safely isolated from the environment. The attempt to build such a repository at Yucca Mountain has come to no avail up to the present time. In fact, no functioning repositories now exist anywhere in the world. Proposed repository sites have all been criticized for failing to meet the criteria for geological stability, and the projected opening date for these facilities keeps receding as new problems are encountered. The Department of Energy currently assumes that the high-level waste repository at Yucca Mountain in Nevada] will not be available is the foreseeable future. There is no guarantee that this proposed site will ever be ready as a suitable repository. If Yucca Mountain passes the assessment, DOE will then apply to the Nuclear Regulatory Commission for a license to build and operate a repository designed to hold 70,000 tons of waste. Federal regulations require that the repository keep waste from reaching the environment for 10,000 years.

Chapter 8. Nuclear Energy

Federal geologists discovered several previously unknown faults slicing through Yucca Mountain. The implications of these findings remain unclear. Yet project opponents say the presence of such faults will hamper plans for storing the nation's most dangerous refuse within Yucca Mountain. Geologists have long known that at least one fault cuts through this region. However the U.S. Geological Survey started an extremely fine-scale mapping program. It is now known that the principal fault running through the proposed repository area actually has several parallel strands extending across a width of 215 meters. This feature, called the Ghost Dance fault zone, runs north-south, as do most of the faults around Yucca Mountain. Recently, a different zone of parallel faults have been discovered that trend northwest-southeast through the planned waste site. The presence of the Sundance fault could—under certain conditions—lead DOE to limit the size of the repository. DOE will have to decide whether to work around these faults. At present, it is not known whether the Sundance and Ghost Dance faults have generated earthquakes within the last several million years. But even if these structures are not active, they may still threaten the storage facility. Because fractured rocks fill these faults, they could provide a path for groundwater to reach the repository, potentially speeding up the rate at which radionuclides leak into the environment. Faults could have the opposite effect, however, if they contain natural mineral cement that inhibits water flow. The basic question is whether or not any of the faults are barriers to the transport of fluids and gases or whether they are conduits.

One unique aspect of the danger associated with radioactive wastes derives not only from the extreme toxicity of these materials but also from the great longevity of many of their isotopes. Cesium-137 and strontium-90 have half-lives of 30 and 28 years, respectively, and must be isolated for 300-600 years. Plutonium-239 has a half-life of 24,400 years and must be isolated for 240,000-480,000 years. Recorded human history covers 6,000 years. The huge quantities of highly toxic radioactive materials must be safeguarded for thousands of years—more than the present age of the pyramids which are the oldest durable monuments of man.

Transuranic (TRU) waste

Transuranic (TRU) waste is defined as waste that is contaminated with alpha-emitting radionuclides of atomic number greater than 92 and half-lives greater than 20 years (primarily plutonium) in concentrations greater than 100 nanocuries per gram (nCi/g). That is, Transuranic (TRU) waste is radioactive waste that is not HLW but contains more than 100 nanocuries per gram of elements heavier than uranium (the elements with atomic number higher than 92). Most TRU waste in the United States is the product of defense reprocessing. TRU includes wastes, whether low level or high, that contain elements heavier than uranium (hence the name "transuranic") in amounts greater than 100 nanocuries per gram. It is destined for geologic repository disposal. It decays primarily through the emission of alpha particles, which have short range, generate little heat, and are readily blocked. Radioactive waste with less than 100 nCi/g of TRU is, in effect, low level waste and disposed of in near-surface facilities, as appropriate. TRU waste must meet rigorous acceptance criteria (and sometimes processing) before it is ready for transportation and repository disposal.

Transuranic elements do not occur naturally; these wastes are produced mainly, though not exclusively, in the military nuclear programs. They are separated during the reprocessing of spent reactor fuel from the Navy and during reprocessing to obtain weapons-grade plutonium produced in special "production" reactors. Most TRU waste is separated and stored at the Idaho National Engineering Laboratory near Idaho Falls and at the Hanford Reservation in Washington; both of these federal facilities have reprocessing programs for military materials. Present policy requires that the TRU waste be incorporated into borosilicate glass, encapsulated in canisters of stainless steel ten feet high and one foot in diameter and weighing 1,000 pounds, and emplaced in deep geologic storage. A large cavern has been excavated for this purpose in bedded salt 2,100 feet below the surface near Carlsbad, New Mexico. Called the Waste isolation Pilot Program (WIPP), all of the preparatory work has long been completed, but opponents have been successful in preventing the use of the facility. The Waste

Chapter 8. Nuclear Energy

Isolation Pilot Project (WIPP) was found to be unable to withstand the thermal stress from spent fuel and will serve instead as the TRU waste repository for the United States. WIPP is scheduled to begin receiving waste in the early 1990s. Because a few drops of water (water of crystallization from the salt under pressure) have been found dripping from the ceiling of some of the mile-long drifts, anti-nuclear activists have raised alarms that the waste may be dissolved in groundwater and reach the surface and get into food. How the moisture could destroy the waterproof stainless steel jacket and dissolve borosilicate glass is not explained by the anti-nuclear forces. Nor is it revealed that if all the water now flowing through the ground in that part of New Mexico were diverted through the salt formation, it would take one million years to wash all the salt out of the repository. It's also useful to remember that glass artifacts from ancient Babylon known to be at least 3,000 years old have spent that time in flowing river water without being eroded. Glass does not dissolve in water—even in salty water— and can with high probability maintain its integrity over millions of years.

Although all scientific evidence supports the safety of long-term (many centuries) isolation deep underground in our radioactive earth, opposition continues. Yet, if all nuclear waste were put into the ground, that would increase the amount of radioactivity in the top 2,000 feet of soil in the United States by only one part in ten million. Even so, this may not be the best way to handle TRU waste. An alternative has been suggested. A new substance named CMPO is capable of selectively isolating transuranics from the rest of the nuclear waste. By removing the TRUs, the remaining waste falls under the definition of low level waste and hence is easier to handle and manage economically. The removed TRUs are by a factor of from 100 to 1,000 times less in volume and can be solidified and vitrified.

Low-level waste (LLW)

By-product material is any radioactive material, except fissile nuclides, that is produced as waste during plutonium production or fabrication. Uranium mining and mill tailings are the pulverized rock and leachate from uranium mining and milling operations. LLW includes everything that is not included in one of the other four categories. The distinguishing feature of LLW is that it contains virtually no alpha emitters. LLW is not necessarily less radioactive than HLW, and may even have a higher specific activity (Curies/gm). Low level wastes account for only one percent of the radioactivity—but 99 percent of the volume—of all radioactive wastes. Low-level wastes may include a wide range of materials, many of which are intensely radioactive. A typical 55-gallon drum of medical low-level waste will contain on the order of 10 millicuries. A typical drum of waste water from a nuclear reactor may contain 10 curies, and a drum of irradiated reactor components may contain 200 to 1,000 curies. Commercial power reactors produce most of all low-level radioactive wastes, whether measured by volume or by total curies. The NRC has designated several classes of LLW:

> **Class A** contains only short-lived radionuclides or extremely low concentrations of longer-lived radionuclides and must be chemically stable. Class A waste may be disposed of in landfills without particular stabilization as long as it is not mixed with other hazardous or flammable waste. Most trash is Class A waste.
>
> **Class B** contains higher levels of radioactivity, must be physically stabilized before transportation or disposal, and cannot contain free liquid.
>
> **Class C** is waste that will not decay to acceptable levels in 100 years and must be isolated from the environment for 300 years or more. Power plant LLW is in this category.

Low level wastes are those with an activity below 0.01 curies per kilogram. That's about one billion times less than the radioactivity in high level wastes. LLW comes mainly from industrial activities—for example, nondestructive testing, which, together with medical waste and that resulting from academic and research uses, accounts for about 43 percent, by volume, of radioactive waste. The remaining 56 percent (by volume) comes from nuclear power plants, and it includes the solidified radioactive nuclides removed from cooling

Chapter 8. Nuclear Energy 217

water, protective clothing, and cleanup materials. Regardless of source, all low level wastes are solid or must be solidified and appropriately packaged before being sent to a storage site. Three such sites have been in use for many years, one in South Carolina, another in Nevada, and a third in Washington State. In each case, the containers of LLW are inspected, monitored, and placed in shallow burial pits or trenches. New LLW disposal sites are being constructed under a law that allows states to form cooperative compacts to take care of their low level waste and hence reduce transportation.

Low-level wastes have been stored at a number of waste dumps. The sites that have been studied most extensively are those associated with the nuclear weapons program. All of these have experienced major breaches of containment, with significant (in some cases, massive) contamination of surrounding areas. Current federal law requires states, acting individually or in cooperation with other states, to build a new generation of "low-level" waste site dumps, but there is no reason to believe that these will contain their wastes better than existing dumps.

Environmental Impact of Nuclear Weapons

Nuclear weapons production and testing for nearly 50 years has produced significant contamination with potentially disastrous human health consequences. Radionuclides have been dispersed worldwide. The approximately 1,400 underground nuclear tests conducted around the world from 1957 through 1989 took place at sites in the US, Russia, the Pacific islands, Algeria, China, and India. The radionuclides underground due to these tests amount to 5 million curies of Sr-90, 8 million curies of Cs-137, and a quarter million curies of Pu-329. The 423 atmospheric tests conducted from 1945 to 1980 produced a global inventory of long-lived radionuclides of about 12 million curies of Sr-90, 20 million curies of Cs-137, 10 million curies of carbon-14, and a quarter million curies of Pu-239. These radionuclides have been dispersed around the world, with the majority of the fallout in the northern hemisphere. The result is that all living organisms, including humans, have incorporated plutonium (Pu) and other man-made radionuclides with carcinogenic, teratogenic, and mutagenic properties into their tissues. Some of these substances will remain in the cells of living organisms for very long periods of time. The Pu-239 isotope of plutonium has a half-life of approximately 24,000 years. Weapons sites are among the most polluted areas on Earth. Little is known about the degree of exposure of workers to radionuclides and hazardous chemicals at these sites. The major ones in the United States are:

Hanford

The Hanford Reservation on the Columbia River in southeastern Washington used to the largest producer of weapons-grade plutonium. In 1943 in Hanford, Wash., President Franklin D. Roosevelt gave 1,500 residents 30 days' notice to move and brought in 95,000 workers to secretly produce plutonium for an atom bomb. The plutonium produced there was used in the Trinity Test, the first nuclear detonation, and in Nagasaki. At its peak in 1964, nine plutonium production reactors were operating at Hanford.

The waste ended up in the ground, in the Columbia River and in tanks that leak to this day. Groundwater contamination is widespread, with tritium and nitrates found in plumes totaling 122 square miles. Carbon tetrachloride, chromium, cyanide, trichlorethylene, uranium, cobalt (Co-60), technetium (Tc-99), iodine (I-129), and strontium (Sr-90) are in groundwater at levels that exceed current drinking-water standards. Past disposal practices have contaminated surface-water sediments. Since 1944, 760 billion liters of contaminated water (enough to create a 12-meter-deep lake the size of Manhattan) have entered groundwater and Columbia River; 4.5 million liters of high-level radioactive waste leaked from underground tanks. Officials knowingly and sometimes deliberately exposed the public to large amounts of airborne radiation during the period from 1943 to 1956.

Idaho National Engineering Laboratory

Chapter 8. Nuclear Energy

This facility in southeastern Idaho 22 miles from the city of Idaho Falls used to reprocess reactor fuel to recover uranium (U-235). Groundwater is contaminated with carbon tetrachloride, tricholorethylene, I-129, Pu-238, Pu-239, Sr-90, and tritium.

Fernald

The Fernald feed materials Production Center in Fernald, Ohio converts uranium into metal ingots.   This facility, located 20 miles northwest of Cincinnati, produces the uranium metal needed for nuclear weapons.   Since plant's opening, at least 205 tons of uranium oxide (and perhaps 6 times as much as that) has been released into the air. Offsite surface and ground-water are contaminated with uranium, cesium, thorium. High levels of radon gas have been emitted. Contaminated groundwater poses the greatest threat to human health; private, community, and industrial drinking-water wells are all contaminated. In addition, surface water and fresh water sediment contamination have been confirmed both on and off site. Soil on the site is also contaminated. Major hazardous substances released into the environment include a number of radionuclides, heavy metals, inorganic chemicals, volatile organic compounds (VOCs), and asbestos. Since 1951, anywhere from 394,000 to 3,100,000 pounds of uranium dust were intentionally released into the air around the facility.

Nevada Test Site

The Nevada Test Site, located 65 miles northwest of Las Vegas, has conducted 100 atmospheric and 714 underground nuclear tests. Environmental contamination has been assessed only in a preliminary fashion. The groundwater has been documented to contain krypton (Kr-85), chlorine (Cl-36), ruthenium (Ru-106), Tc-99, and I-129. Current studies have indicated that it has not migrated off the site. However, numerous studies have been done on downwind communities. The relative risk of leukemia for all ages was 1.72 times the normal risk, and the relative risk of acute leukemia for ages 0-19 was 7.82 times the normal risk.   A dose-reconstruction study of fallout from the site showed that the average external gamma dose per person was about 480 millirem.

Oak Ridge Reservation

This facility, in eastern Tennessee 15 miles from Knoxville, produced nuclear weapons components. Since 1943, thousands of pounds of uranium were emitted into atmosphere. Groundwater, surface water, and soil are contaminated by a large number of radionuclides, heavy metals, and VOCs. Radioactive and hazardous wastes have severely polluted local streams flowing into the Clinch River. Sediment of the Clinch River and several of its tributaries is contaminated with cesium (Cs137), Pu-239, and Pu-240. Watts Bar Reservoir, a recreational lake, is contaminated with at least 175,000 tons of mercury and cesium.

Rocky Flats Plant

The Rocky Flats plant, 16 miles from Denver, is a metal foundry that manufactures the plutonium "triggers" for new warheads and extracts plutonium from retired ones. Large amounts of solvents are used in these processes. More than 7,500 chemicals have been identified at the plant site. Since 1952, 200 fires have contaminated the Denver region with unknown amount of plutonium. Strontium, cesium, and cancer-causing chemicals leaked into underground water. Two creeks run through the property into Great Western Reservoir and Standley Lake, both public drinking-water reservoirs. The sediment of both reservoirs is contaminated with plutonium (Pu-239, Pu-240) and with tritium; the latter was accidentally released into the Great Western Reservoir in 1973. Besides ongoing plutonium emissions from daily operations, there have been three accidents that have released plutonium into the air over the Denver metropolitan area. The first was an explosion and fire in 1957. The second involved windstorms that dispersed soil contaminated with radioactive lathe oil that had leaked from corroded 55-gallon storage drums. The third incident was a smaller fire that breached the building in which it had started. Soil, sediment, groundwater, and surface water have all been contaminated with a large number of

Chapter 8. Nuclear Energy

radionuclides, heavy metals, inorganic chemicals, VOCs, PCBs, and asbestos.

Savannah River Plant

This facility, on the Savannah River, 13 miles south of Aiken, South Carolina, produces tritium and plutonium for nuclear weapons. Groundwater contamination with multiple radionuclides, inorganic chemicals, and VOCs has been found on site only. Radioactive substances and chemicals found in the Tuscaloosa aquifer are at levels 400 times greater than government considers safe. The plant released millions of curies of tritium gas into atmosphere since 1954.

Russia

In 1945, Russia built the Mayak Chemical Complex in the Ural Mountains to produce plutonium. By 1949 they were testing bombs in Kazakhstan. Mayak (also known as Chelyabinsk-40) appeared on no maps and formed the backbone of the Russian nuclear weapons complex. At Mayak waste was dumped directly in the Techa River, in lakes and into tanks that still leak. Lake Karachay is considered the world's most radioactive site. A person standing on its shores would receive a lethal dose of radiation in one hour. In 1957, one storage tank exploded and hundreds of thousands were exposed to the fallout. The CIA knew about the disaster but kept it quiet to avoid a closer inspection of US bomb-making facilities.

Other Sources of Radioactive Waste

The increasingly widespread use of radioisotopes in research, medicine, and industry has created a lengthy list of potential sources of other radioactive waste. Sources range from a large number of laboratories using small quantities (a few isotopes) to large medical and research laboratories where many different isotopes are produced, used, and wasted in large volumes.

Liquid scintillation counting has become an important biomedical tool and produces large volumes of waste organic solvents such as toluene, which have a low specific activity. The long-lived contaminants of liquid scintillators are tritium and C^{14} for the most part. These wastes are characteristic of mixed wastes: mixtures of hazardous and radioactive waste. Their chemical nature as well as their radioactivity poses disposal problems.

Naturally occurring radionuclides and those inadvertently released in times past can also pose threats to public health. An outstanding example is the buildup of Rn222 in homes and commercial buildings with restricted air circulation. Rn222 is a member of the uranium decay series, and is thus found ubiquitously in rock. Chemically, Rn222 is an inert gas like helium. When uranium-bearing minerals are crushed or machined, Rn222 is released; there is even a steady release of Rn222 from rock outcrops. Buildings that are insulated to prevent convective heat loss often have too little air circulation to keep the interior purged of Rn222. Although Rn222 has a short half-life, it decays to much longer-lived metallic radionuclides.

Coal combustion, copper mining, and phosphate mining release isotopes of uranium and thorium into the environment. K40, C14 and H3 are found in many foodstuffs. Although atmospheric nuclear testing was discontinued many years ago, fallout from past tests continues to enter the terrestrial environment.

A radionuclide is an environmental pollutant because of a combination of properties:

Radioactive Waste Management

From an air transport standpoint, all nuclear waste handling and disposal facilities fall into two general categories:

> (1) those that have a planned and predictable discharge to the atmosphere and

> (2) those where any discharge would be purely accidental.

Boiling-water reactors, fuel processing plants, and a large number of atomic experimental systems— particularly military—fit into the first category and on occasion also into the second category. Good examples of the second category are well contained pressurized water reactors and long-term storage facilities.   The atmospheric

Chapter 8. Nuclear Energy

pathways from both types of discharges have received a great deal of attention because of the analogies between radioactive emission, sulfur dioxides, and inert gases generated by automobiles and smoke stacks across the nation. Airborne radioactivity is available for inhalation and/or disposition on water supplies or food chain land and thus impacts public health. Much of the discussion about air pollution and meteorology is also applicable to the study of radioactive particles.

Water transport occurs whenever radionuclides in surface or subsurface soil erode or leach into a watercourse or whenever fallout occurs from the atmosphere. The radioactive waste is transferred from its solid or liquid state into groundwater or surface water supplies where the radionuclides can enter the human food chain or drinking water. Little is known, however, about the leachability of radionuclides, and only gross estimates can be made about surface water transport and disposition. As with any such radioactive material transport equations, the half-life of the isotopes must be considered.

Radioactive isotopes also move from the land into the human food chain. Plant uptake mechanisms have virtually no way of screening an element that is radioactive; thus plants generally collect a deposit of radionuclides in their structure if the radionuclides are present with the water, phosphorus, nitrogen, and trace elements necessary for plant growth. Rain and wind also act to deposit the hazardous material on the roots and leaves of the plants, all to be passed along to whatever or whomever eats the plants. Cows eating contaminated corn, for example, can pass radioactivity along to milk drinkers.

Transuranic wastes last for long periods of time within the environment, but for the most part are strongly held by soil particles. They are not easily translocated through most food chains although some concentration does pass through certain aquatic food systems. In addition, they pose only a slight biological hazard to humans because adsorption in the gastrointestinal track is limited. The greater potential hazard from TRU wastes is from inhalation of dust particles containing the materials since a large fraction is generally retained in the human lung.

In summary, the variety of chemical characteristics displayed by radioactive materials allows them to be transported through the environment by a number of different pathways, making the management of such wastes especially troublesome.

The objective of the environmental engineer is to prevent the introduction of radioactive materials into the biosphere during the effective lifetime (about 20 half-lives) of these materials. Control of the potential direct impact on the human environment is necessary but not sufficient because radionuclides can be transmitted through water, air, and land pathways for many years and in some cases for many generations. Some radionuclides that are currently treated as waste may be retrieved and recycled by future reprocessing, but such recycling creates its own waste stream. Much radioactive waste is not amenable to recycling. It can be treated only by isolating it from pathways to the human food chain and environment until its radioactivity no longer poses a threat. Isolation requirements differ for the different classes of radioactive waste. With the given half-lives of many radioisotopes, it is difficult to imagine any technology that truly offers ultimate disposal for these wastes; thus we will think in terms of long-term storage: 10, 100, or 10,000 years or longer.

High-Level Radioactive Waste

The following options have been considered for long-term disposal of HLW:

> 1. Land disposal. Burial in very deep holes. Burial on an inaccessible island. Deep (mined) geologic disposal (appears to be the only available option). Liquid injection into geologic formations. Rock melting
>
> 2. Sub-seabed disposal (option has been discontinued because the waste could never be retrieved)
>
> 3. Disposal in polar ice sheets

Chapter 8. Nuclear Energy

4. Disposal in space

5. Transmutation into shorter-lived or stable nuclides

In order to store radioactive waste with a reasonable degree of assurance that it will not be dispersed into the environment, a three-stage barrier has been proposed. The first barrier would be provided by the waste form itself: The optimum waste form would be radioactive material dispersed in a glass matrix or vitrified. The second barrier would be provided by an engineered system, which includes the waste packaging. The third barrier would be the geological (rock) matrix.

The word vitrify refers to converting (something) into glass or a glasslike substance, typically by exposure to heat. The option of vitrifying nuclear waste is a feasible solution but it does present problems. Vitrification is planned for defense HLW. Since commercial spent fuel is not reprocessed, however, it will be stored in a geological repository in the form in which it leaves the reactor, as spent fuel rods, which are combined in heavy stainless steel casks in the engineered barrier system.

When the fissile uranium in a fuel rod, in a reactor is about 75 percent used up, the rods are ejected into a very large pool of water where they remain until the short-lived nuclides have decayed completely and the rods are thermally cooler. Initially, this phase was intended to be about six months long, but the spent fuel has remained in storage pools as long as ten years in some cases because there is no repository for it. After sufficient decay and cooling, the spent fuel will be loaded into casks and emplaced in a repository.

Investigations for a geological repository began in 1972. Salt is the most thoroughly investigated substance because:

> Salt mining technology is well developed and storage sites can be constructed.

Salt deposits tend to have a **high plasticity** and thus have a tendency to seal themselves if fractures are created by major movements in the earth's crust.

Salt deposits have low permeability and are essentially sealed from groundwater and surface water supplies.

Salt has a **high thermal conductivity**, which helps dissipate heat that builds in waste containers.

Salt formations have a **high structural strength** with the ability to withstand effects of heat and radiation.

Salt has some disadvantages, however. No salt deposit is free of brine inclusions, and these tend to migrate toward heat sources, which the emplaced radioactive waste would provide. Salt also contains some fossil water.

The Nuclear Waste Policy Act of 1982 mandated two geological repositories and further mandated geological diversity since the behavior of any single medium over such long time periods can only be guessed at. This law was amended in 1987 to postpone investigation of a second repository indefinitely. Three rock types—salt, volcanic tuff, and basalt—were being investigated for suitability for the first repository but only volcanic tuff is presently under study. Except for salt, the media investigated are all hard rock, which cracks and fissures under thermal and chemical stress. Radioactive emissions may even assist in the cracking. Repository integrity thus depends on keeping radionuclides out of ground water aquifers if any are present; this poses potential problems for the hard rock repositories.

There are now about 100 operating nuclear electric generating plants in the United States. Only very few new ones are ordered or designed. Even if the entire nuclear industry were to be shut down, however, the waste would still be with us. Many think that the nuclear industry should be shut down entirely. As nuclear plants finish their useful lives of 30-40 years, they should not be replaced. In this context, we must realize that 20 percent of the electricity in

Chapter 8. Nuclear Energy

the United States today is nuclear-generated. It is unlikely that we will be able to replace this generating capacity with solar, wind, and geothermally generated electricity in the next 30 years, if indeed we can do it at all. Energy conservation can perhaps increase supplies by 5 percent. The favored future option for electric generation is natural gas, which releases carbon dioxide. Thorough and objective comparison of the environmental and public health costs and benefits of the various options must be made before making any decisions that apply unilaterally to nuclear power. It is wise to remember that generating electricity, no matter how it is done, has environmental consequences.

Nuclear Testing

The United States and Russia, in carrying out nearly 2,000 nuclear tests, exposed their own people with radiation and poisoned their land and water in the name of national security. The tests have stopped but the nuclear waste, tons of which are improperly stored, will last for centuries. In the US there are more than 45,300 military and industrial sites contaminated and potentially contaminated with radioactivity, and a similar situation exists in Russia.

The two superpowers acted with disregard for their own people and did it in a strikingly similar fashion. The United States tested bombs and marched its soldiers through the fallout in Nevada and the Marshall Islands. The Russians did the same near the village of Totskoye in the Ural Mountains. the legacy is cancer, birth defects, and weakened immunity. According to Department of Energy documents, from the 1940s through the 1970s, the United States used its citizens as guinea pigs for radiation experiments, without their consent. From 1945 to 1947, 18 patients in hospitals throughout the country were injected with plutonium to measure how much their bodies would retain. In Semipalatinsk, Kazakhstan, medical personnel collect deformed infants and fetuses. There are some of the 200,000 US Atomic Veterans - Americans exposed to radiation at Hiroshima, Nagasaki, or at nuclear test sites. There are many stories of miscarriages and birth defects, cancers and eye cataracts. Compensation depends on proof of exposure and type of cancer. More than 15000 radiation-related claims had been filed.

Between 1951 and 1992 the United States exploded 925 bombs at the Nevada Test Site. Of those tests, 123 sent radioactivity off-site according to DOE records. The US government estimates the tests released over 12 billion curies of radiation into the air. The Chernobyl explosion released an estimated 81 million curies. They assured the downwind people that everything was going to be fine. One secret Atomic Energy Commission memo called the sparsely populated downwind communities a "low-use segment of the population." Sheep began dying and people got cancer. A study of high fallout areas downwind in Utah found childhood leukemia rates 2.5 times the national average. Skin cancers, eye damage, Down's syndrome, and cancer are commonplace. Downwind of the Russian test site near Semipalatinsk, Kazakhstan the story is the same. Three major cancer studies show up to nearly 40 percent more cancers among downwinders. Half the Kazakhstan downwinders studied by the Commission of People's Deputies have weakened immunity. Life expectancy is down. Birth defects are up. Until recently, however, doctors in the area couldn't report high death rates. If the death rate was very high, the minister of health punished the local physician. The Russians conducted more than the 470 tests there.

At Hanford on December 3, 1949, one ton of green uranium was processed, releasing massive amounts of radioactive iodine-131 and xenon-133. Internal memos reveal that the release was deliberate, that the dangers were known and the consequences carefully monitored. In other words, it was an experiment. Within a couple of weeks people's hair started coming out by the handfuls and people got the worst cases of hypothyroidism that doctors had ever seen. Now the cleanup of Hanford is a huge task. Many billions of dollars have been and more will be spent. Some of the highly radioactive waste just sits uncovered like a swimming pool.

At Mayak the same situation exists. Three major accidents occurred at Russia's largest plutonium production center, collectively exposing a half-million people. Now they are dismantling nuclear warheads and storing the waste improperly. The situation there is practically impossible. They have no time to cope with old problems and they have new ones. It's not safe. Many of the workers in the secret city used to get high pay and goods not found in Moscow. Now they feel forgotten, It is possible that plutonium is disappearing. Russia had more than 900,000 people in its nuclear weapons industry. Much of their legacy remains secret. At times, the governments volunteer the information, as the US did when it released information on 204 previously unannounced secret nuclear tests. The Paducah Gaseous Diffusion Plant used to enrich uranium for use in nuclear weapons. It dumped 60,000 pounds of uranium into local creeks and 130,000 into the atmosphere.

Plutonium

Plutonium is not "the most toxic substance known to man." However, plutonium is toxic, but not highly toxic in the ordinary sense, that is, not chemically toxic. In fact, several grams of plutonium would have to be ingested to be fatal in the usual sense of a poison that produces death in a relatively short period of time. On the other hand, plutonium is highly **radio toxic;** that is, toxic because of its radioactivity. Since plutonium's radioactivity is the emission of alpha particles that cannot even penetrate the skin, plutonium must be swallowed or inhaled to be hazardous because of its radioactivity. Plutonium on the skin is not readily absorbed and can be washed off with no residual effects. So, the principal hazard from plutonium is to workers rather than to the general public.

If plutonium enters the body in a form that dissolves in body fluids (it is normally insoluble), some would be transported to the bone and liver, where the alpha-particle bombardment could eventually result in the development of cancer. If plutonium is inhaled into the lungs in a form that will not dissolve in body fluids, some will remain in the lungs and eventually may result in lung cancer. For inhalation doses in the range of the permissible lung burdens or less, the latent period for the development of lung cancers is 15 to 45 years. The chance of developing cancer depends on the amount of plutonium inhaled or swallowed. The incubation time also depends upon the amount of plutonium taken into the body. Large amounts of plutonium are more likely to cause cancer in a shorter period of time than small amounts. Similar results were observed in workers who swallowed or inhaled radium during the 1920's. Plutonium is toxic in a manner similar to that of radium, but, weight for weight, the most common form of plutonium (plutonium-239) is only one-sixth as toxic as the most common form of radium (radium-226). However, the exposure to plutonium should be minimal because of the safety and containment practices and widespread knowledge of the hazard, in contrast to the virtual absence of such practices in the 1920's due to the ignorance of the hazards of radium.

While one single particle might cause cancer, the chances that it will actually do so in a human lifetime are small enough so that it has not yet been observed in humans. A larger quantity of plutonium is, of course, more likely to cause cancer, since the radiation dose is larger.

All work on the plutonium-bearing materials is performed in glove boxes or with automated machines. Glove boxes are nearly gas-tight boxes with a window for viewing the pieces being worked. The boxes are equipped with leak-tight arm-length gloves through which the workers extend their arms into the boxes to handle the work pieces. The insides of the boxes are maintained at a lower pressure than the room air so that any leakage or holes in the gloves will result in in leakage of air to assure that no plutonium or plutonium-bearing dust escapes the closed system. The ventilation of the glove box air passes through a series of high efficiency filters that remove plutonium being carried in it. Sensitive detectors monitor the flow to detect any failures in the filter system. Workers are checked for plutonium on their hands and feet as they enter and leave plutonium work areas to detect contamination and to prevent its spread. After the plutonium-bearing materials are encapsulated (e.g., in a fuel element), the outer surfaces are decontaminated before the finished items are returned to an unprotected environment. Companies handling plutonium are required to protect workers as a condition of their licenses from the U.S. Nuclear Regulatory Commission.

Chapter 8. Nuclear Energy

Not in the ordinary sense of antidotes. If swallowed, only a very small fraction (typically 0.005%) of the plutonium is absorbed in the intestinal tract. If inhaled, the lungs expel the larger particles; if in soluble form, the retained material passes into the blood stream; if in insoluble form, the smaller particles tend to pass into the Lymphatic system of the body, and intermediate size particles tend to remain in the lungs and may cause cancer after long residence times. However, pulmonary irrigation can be used to assist the lungs in ejecting particles. Once absorbed in the body the excretion rate of plutonium can be accelerated somewhat by the use of medicines (chelating agents).

The U.S. Transuranium Registry has been established to identify people who have swallowed or inhaled transuranic materials such as plutonium, to monitor their health for the remainder of their lives, and to accumulate and analyze the statistical data that are developed. This is believed to be the best and quickest method of determining the human effects of plutonium poisoning. So far, however, little data have been accumulated as no adverse health effects have been observed. However, this is a voluntary health program and participation by past plutonium workers has been low.

Dispersion of plutonium in air or the threat of it could be used for blackmail or terrorism if the public is led to believe that its consequences would be disastrous. The dispersion of plutonium is not going to kill anyone in a short period of time and therefore its effectiveness for a terrorist is minimal. Furthermore, the results of plutonium dispersion have been estimated to be small (one fatality for each 15 grams dispersed without warning or 150 grams dispersed with warning). The fatalities would result from cancer 15 to 45 years later.

However, plutonium is not easily diverted for such use. Plutonium is carefully controlled and accounted for, and people working with it will be screened. It would have to be stolen and its users would be taking grave risks unless they were familiar with its hazards, knew the techniques, and had the equipment for handling it. Plutonium is a valuable material as well as a dangerous and hazardous one and is therefore, protected against loss or theft. In 1975 the measures for physical protection of plutonium were greatly increased.

By guarding the plants and vehicles in which it is stored, handled, and shipped. Theft of small quantities at a time is prevented by examining workers with detection equipment in a manner similar to that employed at airports to detect weapons that could be used in hijacking.

Yes. Almost all incidents are the result of weapons production and uses and are not related to nuclear power production. It is estimated that 6.5 tons (5900 kilograms) of plutonium have been injected into the atmosphere by nuclear-weapons and weapons test explosions prior to

Four plutonium bombs were accidentally dropped near Palomares, Spain, in 1966. The non-nuclear (chemical explosive) part of two of the bombs exploded thereby spreading plutonium over a few hundred acres. Some of the land was decontaminated by removal of the topsoil and shipping it to the U.S. for further treatment. In other areas with very low concentrations, the plutonium was further diluted by plowing. No aftereffects have been observed after ten years.

At the Rocky Flats, Colorado Weapons Fabrication Plant, plutonium was released into the atmosphere by fires in 1957, 1965 and 1969 (less than 1 gram). Also between 1958 and 1968, 86 to 1630 grams leaked from drums onto the ground nearby and was dispersed by wind in an area adjacent to the plant. About 80 grams is under an asphalt pad in the plant area.

At Thule, Greenland a military aircraft crash and fire dispersed plutonium into the environment. Plutonium was recovered from the packed snow, ice, and plant debris. Only a small percentage of the plutonium escaped as an aerosol. Accumulations of plutonium from process waste streams were found in a Hanford, Washington trench and in the Erie Canal outside of Mound Laboratory in Miamisburg, Ohio. Most plutonium generated in nuclear power production, as of

Chapter 8. Nuclear Energy

early 1976, is being stored in spent fuel elements awaiting fuel reprocessing. After the fuel is reprocessed, the recovered plutonium will be recycled in newly fabricated fuel elements and will be consumed by fissioning in the production of more power. Plutonium bred in breeder reactors will also be consumed by recycling.

Since the 1950's, production and processing of plutonium for nuclear weapons in large quantities (many tons per year) were conducted without any significant health consequences. It has been charged that the best controls will not be good enough to protect the public. Although the primary group at risk consists of those occupationally exposed, the record shows that even under the early conditions of weapons production, both the public's and the worker's health and welfare were adequately protected. While there were some serious difficulties that have been discussed in this section, the far better methods and controls used today adequately protect both workers and public from plutonium.

Plutonium used in reactors is mostly plutonium-239. Plutonium with a large fraction of the isotope plutonium-238 (half-life of 89 years), which is not fissionable, is used as radioisotope heat sources for thermoelectric power generators used in space applications and for nuclear cardiac pacemakers, a small electronic device implanted in the body to provide timed electric impulses to the hearts of people who have defective natural mechanisms. Plutonium-238 is also being considered as a power source for the artificial heart.

Fuel Reprocessing

The principal reason for reprocessing the spent fuel is to recover valuable unused uranium and plutonium from the spent fuel elements. In addition, the fission product wastes, which are the product of the uranium fissions that generate energy in the reactor, are concentrated into small volumes for disposal.

Only about two thirds of the fissionable uranium (uranium-235) in a fuel element is consumed before the fuel must be removed from the reactor for economic reasons. Recovery of the uranium will reduce the cost of power production by about 2.5%. It will conserve a valuable natural resource, and will reduce (by 10%) the amount of mining necessary to replace it.

Recovery and recycle of plutonium from light water reactors (LWRs) could produce savings about twice those from recovery of the uranium. Concentration of the fission product wastes from spent fuel elements into small, more easily managed volumes also makes disposal of the wastes more economic. The cost of reprocessing has been estimated to be less than 1% of the power generation costs.

Although the quantities of nuclear waste will increase as nuclear power generation grows, the problems of nuclear waste disposal are not related to the volume or weight of the wastes. On the contrary, energy in nuclear fuel is so concentrated that the amount of waste is very small for the energy obtained. This may be seen by comparing the amount of nuclear waste with the waste from coal-fueled power plants.

All the nuclear wastes to be generated by the entire U.S. nuclear power industry from now until the year 2000 could fit into a cube about 250 feet on a side and of that, the ' high-level wastes" would occupy a cube only about 50 feet on each edge. In contrast, it takes a train of about 33 coal cars per day to remove the ashes from a single 1000-MW coal fueled power plant. Another way of putting it is that the nuclear waste generated in providing all the electricity used by an individual in his entire lifetime is about the same size as 100 aspirin tablets. This is very small volume of the nuclear wastes makes its management, by interim storage or permanent disposal, technically feasible and economically practical.

The nuclear waste disposal problem is one of disposing of the wastes by ensuring that they are isolated from human environment for the very long time it takes until they are at natural background (harmless) levels —600 years to several hundred thousand years, depending upon the degree of processing. In the past, it had been assumed that the waste must be isolated until the longest-lived isotope in the material had decayed for about ten half-lives. Since the separation of uranium, plutonium, and fission products during reprocessing is not perfect, the wastes, as presently processed, contain about 1/2% of the plutonium (half-life, 24,000 years) from the spent fuel, enough to require the wastes to be confined for about 250,000 years. By contrast, the wastes can be made no more hazardous than a natural uranium mine in 600 to 1000 years by reducing the residual valuable plutonium in the wastes from 0.5% to about 0.005%, a factor of 100. Conceptual processes have been developed for further stripping of plutonium from the wastes, but their practicality and economic feasibility must still be demonstrated.

The spent fuel elements are stored (under approximately 20 feet of water in licensed storage pools) either at the nuclear power plants or at the reprocessing plants until reprocessing operations begin. The fuel from LWRs in use today consists of bundles of sealed metal tubes (about 1/2 inch in diameter and 12 feet long) containing pellets of uranium dioxide, each about 1 inch long). In reprocessing, the tubes are chopped into small segments, the fuel pellets are dissolved in strong acid, and the unused uranium, the plutonium, and the fission product wastes are chemically separated. The fission product waste that is of greatest concern is called the "high level waste," because of the high levels of fission product radioactivity and heat generated by the radioactivity, not because of its plutonium content. Initially, the heat from radioactive decay of the fission products is enough to boil the liquid waste solutions; therefore, cooling must be provided for several years.

Nuclear fuel reprocessing plants have been in world-wide operation for many years. In the U.S. there are three government plants currently operating, processing fuel only from military programs and special reactors. Less than 10% of the fission products are radioactive when the fuel is reprocessed, and after the 10-year decay, less than 0.05% of the radioactivity remains. In the early government reprocessing plants, storage was in carbon steel tanks and leaks were anticipated. A spare tank was provided to receive the contents of a tank in case a leak developed. Modern storage tanks are made of stainless steel and are doubly-contained so that any fluid leaking through the first barrier will be retained in the outer tank and not escape to the environment. Leakage of the first container would be detected and the tank contents transferred to a spare tank.

At Hanford, Washington 16 of 151 tanks (containing wastes from military programs) developed leaks at weld areas over a period of 20 years, and some radioactivity escaped into the ground However, no radioactivity, except for trace quantities of tritium, has reached or is expected to reach the water table as shown by a continuous sampling program. Tank failures have resulted in large discharges. The largest leak occurred in 1973 when approximately 115,000 gallons of liquid (out of some 505,000 gallons in the 533,000-gallon tank) escaped before the leak was detected and stopped. However, all of the radioactivity was retained within 90 feet of the tank by the highly retentive soil. The Environmental Protection Agency investigated this leak and concluded that "the leakage of radioactive material from tank 106T does not presently constitute a threat to the public health in the foreseeable future."

Newly generated high-level wastes are being stored only in the modern double-walled tanks while waiting for processing to a solid form. Processes for solidifying the high-level wastes and incorporating them into either glass-base or cement materials have been developed and tested. Both base materials are highly impervious to water and ensure that very little of the potentially hazardous waste would be leached out of the solid and be carried away, even if these materials were inadvertently exposed to ground waters for very long periods of time. The huge volumes of water and the slow flow (less than 100 yards/year) characteristics of ground water would further dilute and provide time for decay to innocuous levels.

Alpha wastes are solids generated in fuel fabrication and reprocessing that are contaminated with low levels of long-lived alpha-emitting transuranic elements (plutonium, americium, curium, etc.) but with such low levels of gamma-radiating materials that they can be handled without shielding. They consist of materials such as paper, cloth, wood, plastic, rubber, glass, ceramics, metals, salts,

Chapter 8. Nuclear Energy

sludges, and filters. The alpha wastes are the principal constituent of the so-called "all other wastes." Alpha wastes are larger in volume than high-level wastes, but are easier to contain since they generate essentially no heat and have very low radioactivity levels. High-level wastes, on the other hand, while having large radioactivity levels and generating heat, are small in volume.

Alpha wastes require the same long confinement times as high-level wastes containing 112% of the plutonium, and therefore must be stored in a Federal repository. However, since they are harmless unless ingested or inhaled (alpha radiation is stopped by a sheet of paper), simple confinement in metal containers protected from the weather is adequate. The National Academy of Sciences recommended that solidified nuclear wastes be permanently disposed of by burial in rock salt formations because these salt formations have not been disrupted by earthquakes for millions of years. This gives a high degree of confidence that they will remain geologically undisturbed for the next million years, which is more than the time required for safe permanent disposal. The salt formations have also been free of circulating ground water for millions of years, being protected from aquifers by impervious rock. Thousands of square miles of suitable salt formations are potentially available in the U.S., and the required area of a few square miles is easily available.

Oak Ridge National Laboratory conducted laboratory tests and an engineering field test in a salt mine near Lyons, Kansas for the AEC and considered using the mine for a Federal nuclear waste repository. Further investigation by the Kansas Geological Survey showed that the Lyons site was unsuitable because of nearby abandoned well holes that penetrated the formation and because of an adjacent water mining operation. The AEC abandoned the Lyons site, but ERDA is investigating other thousands of square miles of available salt beds for suitable salt bed disposal sites, particularly in Kansas and New Mexico. Recently, another prospective site near Carlsbad, New Mexico was selected for intensive investigation.

"Long-term interim storage" means that high-level solidified wastes taken from the storage tanks will be canned in leak-tight canisters and placed in engineered vaults at or near the earth's surface on government-owned sites and monitored continually. These wastes will then be retrievable at any time for transfer to permanent disposal or for additional processing to extract components for useful applications that may be developed. The stored canisters will be monitored to ensure their integrity and the contents will be transferred to new canisters as required. "Permanent disposal" means that the wastes are safely isolated from the environment so that no further surveillance or other efforts are ever required. Several methods under consideration, such as disposal in rock salt formations, disposal in other deep geological formations, disposal by transportation out of earth orbit, a long-range proposal, and transmutation to nonradioactive or short-lived radioactive materials in nuclear fission or fusion reactors. From now through the year 2000, a maximum of two government-owned and operated facilities would be required for the solidified long-lived nuclear wastes, one for interim storage and one for permanent disposal assuming the salt bed method is used. The land area required is only a few tens of acres. A few square miles may be desired to provide for supporting facilities, security areas, and an exclusion area around the sites.

The responsible management of nuclear wastes is not a new burden imposed by nuclear power development, but the extension of one started with the introduction of nuclear weapons in the 1940's and the nuclear Navy in the 1950's. The solidified volume of the military wastes exceeds that expected from all U.S. nuclear power plants to the year 2000 by a factor of 10. Therefore, with or without commercial nuclear power, processes for the responsible management of nuclear wastes must be developed and used. However, as noted above, there are several feasible methods for permanent disposal of nuclear wastes. The problem is to evaluate and devise detailed engineering tests that will assure safe disposal by the best method. Meanwhile, long-term interim storage of nuclear wastes can be used, indefinitely if desired, until a permanent disposal method is selected. Thus, there are a number of ways in

Chapter 8. Nuclear Energy 239

which the wastes might be disposed of, but solutions to such long-term problems must be evaluated and tested on an engineering scale for relatively long times (10 to 30 years).

Spent fuel is accumulating since reactors are operating. Thus, decisions are urgently needed with respect to reprocessing and recycling of uranium and plutonium fuels but not on waste disposal. In the 1950's and 1960's when commercial nuclear power was in development, decisions were made that fuel would be reprocessed and recycled. Planning, commitments, and construction of power plants, reprocessing plants, and waste disposal systems were based on these decisions. Now all of these decisions have become subject to the National Environmental Policy Act of 1969 since they had not yet been fully implemented and therefore must be reaffirmed. Indecision is greatly inhibiting the contribution that nuclear power could make to the U.S. energy requirements in the period from 1980 to 1995. Unless timely decisions are made, spent fuel storage capacity will be exceeded in 1978 unless additional storage space is built or reprocessing begins. Reprocessing, in turn, is being held up by decisions on recycling o~ uranium and plutonium fuel.

On the other hand, decisions are not urgently needed on disposal of high-level, solidified nuclear waste. Solidified waste will not be produced from nuclear power operations until the 1980's, since there will be no reprocessing plants in operation before 1977 and there are cooling periods of two to five years after reprocessing before the solidified waste would be ready for disposal. Also, the accumulated quantities of nuclear waste from commercial nuclear power plants will be too small to justify construction of a repository until the late 1980's. Thus, in terms of the quantity of waste, there is plenty of time to consider alternative methods. However, the controversy and concern about waste disposal has led to demands by some for immediate decisions and Environmental Impact Statements on waste disposal.

A legacy of natural radioactive materials (including alpha emitters) far larger than humans will produce has existed since the creation of the earth and yet humans have successfully progressed to their present state. There is also evidence that a nuclear reactor in nature operated billions of years ago for a period of about 500,000 years in the territory that is now Gabon, Africa. Natural ground water and the uranium isotopic levels present at that time were sufficient to support a chain reaction. The fission product activity has long since decayed, but traces of the plutonium and its radioactive decay products are still evident. The stable fission products and the plutonium decay products appear to have remained localized. The moral question of adding our increment of additional radioactivity cannot be considered as an isolated question but must be viewed along with the consequences of using alternatives to nuclear power. We must consider that to sustain our industry, to feed the current world population, and to satisfy the growing demand for an increased standard of living. Conceivably, we could burn up all of the obtainable fossil fuels in several hundred years. Should we leave future generations the legacy of a world without fossil materials (oil, natural gas, and coal) from which to extract fertilizers, medicines, and plastics because we elected to burn them instead of using nuclear power?" It is unrealistic to expect that humanity can exist on its present scale without leaving a trace of both opportunities and problems for future generations.

Disposition of Excess Weapons Plutonium

With the end of the Cold War, some 100 tons of excess plutonium resulting from the dismantlement of thousands of nuclear weapons are a danger to international security. None of the options yet identified for managing this material can eliminate the danger. All they can do is to reduce the risks. At present the greatest risks, posed by the nuclear weapons and fissile materials, include:

- **Breakup**—emergence of multiple nuclear-armed states where previously there was only one;

- **Breakdown**—erosion of government control over nuclear weapons and materials within a particular state;

- **Breakout**—repudiation of arms reduction agreements and pledges, and reconstruction of a larger nuclear arsenal.

Chapter 8. Nuclear Energy 241

A safe storage of nuclear weapons materials in the short term must be found coupled with new agreements on wide-ranging information exchanges and monitoring. Nuclear weapons or materials must not fall into the hands of unauthorized parties. Actions are required to secure and account for these weapons and materials.

Safeguarded storage is a necessity stretching over decades because all practical methods of disposition are time consuming. For the longer term, the committee urged that options for disposition of plutonium be designed to meet a "spent-fuel standard"— making plutonium withdrawn from weapons as difficult to acquire as the much larger amounts of plutonium in spent fuel from civilian reactors. This should be done as quickly as possible.

The two most promising technologies are vitrification—a way of processing the material into a glass-like substance that would include high-level radioactive wastes—or use of the material as fuel in existing or modified nuclear power plants. Both of these options would embed the plutonium in intensely radioactive waste forms from which it would be difficult to recover for use in weapons. A third option would be to bury the plutonium in boreholes several kilometers beneath the Earth's surface. This last option has been less studied but could turn out to be comparably attractive.

We must rule out a wide array of technologies that had been proposed for plutonium disposition, from underground explosions to launching the material into space. The need to deal with excess weapons plutonium should not drive decisions about the future of nuclear power. Advanced reactors should not be specifically developed or built for transforming weapons plutonium into spent fuel, because that aim can be achieved more rapidly, less expensively, and more surely by using existing or evolutionary reactor types,.

An expanded cooperative program is necessary to reduce security risks, including international assistance in providing accounting and monitoring devices to detect any theft of material that can be used to make weapons. A verified halt in the production of fissile material for weapons, and monitoring of weapons dismantlement, would be key parts of this program. While some weapons will inevitably be kept for reserves, both the United States and Russia should commit a very large part of their nuclear materials from dismantled weapons to non-weapons uses. At present, neither country has decided how much material will never again be needed for weapons. Improved management of materials from weapons dismantlement in the United States and Russia should serve as a steppingstone for improving safeguards and security for all fissile materials worldwide. For the future, new consideration of ways must be given to balance nuclear power against the risk of nuclear proliferation, including continued research on options for near-total elimination of plutonium.

In addition to plutonium used for weapons, there are hundreds of tons of civilian plutonium in spent fuel around the world, a by-product of the nuclear power industry. Another 6070 tons are produced every year. Some of this material is being separated from the spent fuel for use as reactor fuel, although such an approach currently is more costly than using uranium. As of late 1992, some 86 tons of plutonium separated from spent fuel from nuclear reactors were in storage. Although reactor-grade plutonium differs from weapons-grade plutonium, it still can be used for weapons. Even with relatively simple designs such as that used in the Nagasaki weapon—which are within the capabilities of many nations and possibly some subnational groups—nuclear explosives could be constructed [from reactor-grade materials that would have yields of at least one or two kilotons. Steps should be taken to reduce the proliferation risks posed by all of the world's plutonium stocks, military and civilian, separated and unseparated.

www.ingramcontent.com/pod-product-compliance
Lightning Source LLC
Chambersburg PA
CBHW070621220526
45466CB00001B/70